Extreme Conservation

EXTREME conservation

Life at the Edges of the World

JOEL BERGER

The University of Chicago Press
Chicago and London

The University of Chicago Press, Chicago 60637
The University of Chicago Press, Ltd., London

Published 2018
Printed in the United States of America

27 26 25 24 23 22 21 20 19 18 1 2 3 4 5

ISBN-13: 978-0-226-36626-5 (cloth)
ISBN-13: 978-0-226-36643-2 (e-book)
DOI: https://doi.org/10.7208/chicago/9780226366432.001.0001

Library of Congress Cataloging-in-Publication Data

Names: Berger, Joel, author.
Title: Extreme conservation : life at the edges of the world / Joel Berger.
Description: Chicago ; London : The University of Chicago Press, 2018. | Includes index.
Identifiers: LCCN 2018001006 | ISBN 9780226366265 (cloth : alk. paper) | ISBN 9780226366432 (e-book)
Subjects: LCSH: Nature—Effect of human beings on. | Climatic changes.
Classification: LCC GF75 .B474 2018 | DDC 304.2/5—dc23
LC record available at https://lccn.loc.gov/2018001006

♾ This paper meets the requirements of ANSI/NISO Z39.48–1992 (Permanence of Paper).

You've flown for twenty-one time zones to get to this polar island in winter. Two years ago, you did the same. My government let you come. Now, Russian Security Forces detain you.

Do you think geo-politics will affect the way we do science; the way we cooperate? And, you, the way you Americans do conservation?

ALEXANDER GRUZDEV, in conversation
with the author, 2016

Contents

Foreword

Rarely has a book about conservation been better named. This is a book about extreme conservation, written by a scientist who is an extreme conservationist. Unpacking that sentence will be a pleasure.

The obvious connotation of that phrase is purely descriptive. This is a book about conservation in the extreme climates of this world—the high northern climes and the high altitudes at the roof of the world. These are places of high and low temperatures, of rarified air, and little access to water. Water plays a big role in the story, and it is frequently difficult to access, either because there is too little rainfall or because it is locked up in ice and snow. Sometimes ice tsunamis, or *ivu* in the Inupiat vernacular, can overwhelm animals. Water pooled in lakes can freeze but also fracture and break up. Rivers can carry one away or stop movement across the landscape. For the animals in this extreme climate, in addition to the constant struggle to evade predators and getting enough food to survive and reproduce, there is the constant danger of death through misadventure or just bad luck.

The narrative dwells on the species that are adapted to these extreme and distal climes—frequently the goats and sheep, members of the tribe Caprini. Among mammals, these are the climbers and extremists, what George Schaller called the mountain monarchs: chirus, adapted to the high Tibetan plateau; blue sheep, with their suction-like hooves; takin of the high alpine meadows of the Himalayas; and of course the muskoxen, the largest land mammal of the far north, which hold pride of place in this story. The narrative includes saigas, whose lineage now is considered closer to the antelopes than to the goats. And yaks, a member of the Bovini tribe but one adapted to the high altitudes. The lives of these species are played out against a backdrop of the people that live in these places. Some, like the people in Tibet and Mongolia, are pastoralists, and their domestic goats, sheep, and yaks compete with the wild species for space and forage. Some, like those in Beringia, are hunters, and they see muskoxen as competitors for the caribou on which they depend.

This is also a book about adaptation. All of these species have physiological and behavioral adaptations to withstand the extreme. In winter, muskoxen huddle with their herd mates, their short little legs protected by furry skirts. Yaks illustrate the physical convergence. Behavioral adaptations abound: lactating yaks hover near snow patches because of their need for ready available water. Saigas and chirus are constantly on the move. For all, living in groups helps them survive. The elegance of adaptation can be awe inspiring. And of course, the discussion of adaptation has to consider climate change, because it is in these extreme climates that those changes are most pronounced. Not only is their environment changing, but the changes advantage or disadvantage their predators and competitors. There will be winners and losers. And the world of these species is changing so fast that genetic change cannot keep up. Furthermore, these large-bodied species—slow breeders with long generation times—have less capacity to adapt genetically. But their social and behavioral adaptations allow them to respond to change, and much in this book is Joel Berger's teasing apart the "how quickly, where, and when" questions.

Set in this fascinating social and physical milieu, this is a book about Joel Berger, who as a scientist has sought to understand how these species have adapted and how they may adapt in the future. To do so means doing fieldwork in these places, which is a challenge to life and limb. Joel has been attracted to these extreme environments throughout his life. His early work with rhinos in the deserts of Namibia took him to a place of heat and dryness. Now he explores the places of cold. Joel has always had a special focus in his studies of adaptation. Not only does he have the traditional biologist's fascination with the complex docking of an animal's behavior and physiology with its environment, but over the years he has also explored how animals learn to adapt to new challenges or remember how to adapt to old ones.

Joel's work is informed by traditions grounded in behavioral ecology, and his methodology often derives from that training. Joel constantly seeks to get inside the minds of the animals and uses their behavior to understand the selective pressures on the species and how they have adapted to those pressures. Jakob von Uexküll wrote in the early 1900s about the importance of understanding the *umvelt*, or environment, of the animal, seeking to see the world through the eyes of the animals. Few conservationists have taken those instructions so literally and gained so much knowledge from doing so. Dressed as a grizzly or a polar bear, seeking to understand the importance and the changing nature of predation resulting from the loss of sea ice on muskoxen, Joel delicately approaches herds and gauges their response. He has done this before. His classic studies of the fear response of moose and elk to reintroduced wolves in Yellowstone provides deep insights into the capacity of animals to adapt to novelty and to learn fear.

But in these extreme environments, the challenges to personal safety are more extreme. Unable to outrun the muskoxen, his experimental subjects, what is he to do if the adaptive response of the animal is to charge him down? What is he to do if through the windswept snow he runs into the real thing—an actual polar bear. He treats the possible ignominy of it all with humor ("at least mating season is not until June"), but Joel's humor does not detract from his determination to use experiments to explore how animals think.

It is this capacity to get inside the minds of his subjects that gives his work a special poignancy. Few researchers agonize so publicly on the fate of the individual animal, be they orphans, whose muskoxen mothers were decapitated, or radio-collared females, who, when released, cannot rejoin their herds. And when it is his own actions that create the problem, Joel's feelings are especially excruciating. But Joel is also a scientist, so the socially deprived female provides an opportunity to understand how sociality is important to muskoxen. In a previous project, this same opportunism allowed Joel, when faced with dehorned rhinos in Namibia, to understand the adaptive value of the horn to calf survival, a realization that upended conservation practice in that country.

Joel is a field biologist. Being in the field is not a necessary hardship to get information and knowledge but, instead, has its own value. "I need to be in the field." This particular fieldwork is physically tough, and for those of us more comfortable in the tropics, this extreme cold is almost incomprehensible. Joel seems to thrive on discomfort, and social and political difficulties are obstacles to be worked around, ignored, or otherwise avoided. But the sheer grind and the cold can seep into even Joel's bones, and there are moments of reflection that give this book its real humanity.

The title of this book is *Extreme Conservation*, and the reason to obtain the scientific knowledge gained is to put it in the service of conservation. The species studied here and the many others from far beyond have been hard hit by human depredations, in either the distant or the recent past. Muskoxen were extirpated from Alaska soon after whaling ships began plying the waters of Beringia in the

1800s, and present populations are the result of subsequent reintroduction in the 1930s. The same is true of the Russian population on Wrangel Island. Chiru, the Tibetan antelope, were heavily poached for the trade in shahtoosh shawls (made from the fine underfur of chirus) in the late 1900s, though conservation efforts have stabilized the population at the present time. Historically, saiga populations were managed well in the Soviet Union, but following the USSR's breakup, illegal hunting to supply the traditional Chinese market decimated both the Mongolian and the Western species. Takin have always been rare in historical times, and even populations in protected areas today are of unknown status. While domestic yaks are ubiquitous on the high Qinghai-Tibetan plateau, wild yaks have been pushed into the most remote areas. The small remaining population is vulnerable to introgression from domestic animals and to overhunting as the plateau has been opened up by roads.

Conservation models have not been working well on these species in these inaccessible lands. Living in places of such low primary productivity, their densities are low. Populations require big spaces and the ability to move widely across the landscape. Hemmed in by human populations, populations can be isolated and have few degrees of freedom. Domestic livestock compete for pasture, and ubiquitous free-ranging domestic dogs pose constant predatory threats. Wild populations of all of these species are hunted for their meat and body parts. And climate change generates a constantly shifting set of requirements. Joel's work does not provide all the answers, but he frames the possible adaptive responses of the animals and gives us a necessary optimism. The very fact of their survival in these extreme and remote areas gives us some confidence that they will be able to face the latest set of challenges. What we, as humans, need to do is allow them the space to do so.

John G. Robinson
Chief Conservation Officer
Wildlife Conservation Society

At the edges of the planet, the lands are raw. Ice permeates and penetrates. It has done so in the past. It still does. The poles do not own the ice. Its cold curtain also clothes the planet's tallest and most sacred vertical domain–the Roof of the World [i.e., the Tibetan Plateau]. Animals of these once inaccessible dominions share a Pleistocene ancestry. Born of these distant geographies some species remained steadfast. Others navigated a land bridge to a new continent. We humans are among the pioneers that crossed Beringia from Asia into North America. The species we see today are the ones that endured. What of their future at the hands of the world's most profound architects–us–in a seriously altered climate?

Can extreme species adapt?

AUTHOR'S NOTEBOOK

Prologue

Chukchi Sea, Beringia, Alaska—winter 2012.

The tundra is bleak today—minus 18°F. The wind bites.

Beyond the Inupiat village of Kotzebue, the hills disappear as earth, sky, and sea fuse into a cauldron of angry gray. It's not always like this, even in winter dormancy.

Stark and unadulterated, there is a ghostly mysticism. Turquoise flares stream across starry nights with the crackle of aurora borealis. During the few hours of low daylight, one finds the tracks of a hare gliding across a white surface. There are the remnant craters left in snow from where caribou searched for lichens.

Indoors the mood is festive. A rock-hard chunk of some meat is on the floor. For months it remained outside, testimony that nature's wintry refrigeration still works. Two more frozen morsels are set down, the three delicacies that will become our savory dinner. I can't tell whale from walrus or caribou, but Levitt Angutiqouaq, an Inuit hunter over from Greenland, can. The beluga was harvested here in the Chukchi, and the caribou shot on the permafrost. I didn't ask about the walrus.

We sit cross-legged, each with a sharp knife. To start the banquet preparation, we shave thin strips from each frozen mass, then dip them in olive oil mixed with hot sauce. I chew muktuk, whale blubber. It forms an indigestible bolus that I awkwardly remove. The slivers of iced walrus do not make my stomach happy. The caribou is slightly better. Raw and solid is not my cup of tea. But for Mr. Angutiqouaq, this is an ideal repast.

Atop the world, as elsewhere, life depends on filling one's gut by relying on harvested stores. If you are a seal, it's your own body fat you use. If human, it's that fatty seal you crave.

A disquieting desperation pervades the essence of Arctic animals. The accelerating loss of ice signals only one thing—a violent change in the world's cooling system. For the people who rely on nature's northern land and sea bounties, the change is more rapid than when their ancestors colonized these cold barrens. Such alterations are felt far from the top of the world. The Third Pole—the Himalayas and Tibetan Plateau—awakens from deep sleep as it, too, sheds a glaciated coat.

Ice divides. It can unite. It may be quiescent for thousands of years, but there is no denying its geographical influence. Where continental connections dissolved to form islands, new varieties of life were spawned. Ice also forms fortresses. Its long presence has at times rendered Alaska exclusively Asian. Since being severed from North America, Asia has been pure Siberian. Had your grandmother lived to be a hundred, then it was only 130 grandmothers ago when a new gateway opened and allowed the spread of Asian travelers into North America.

Ice has also indelibly shaped the land. The broad steppes of central Asia and lowland grasslands of the Indian subcontinent were separated by the encased Himalayan massif. Species on both sides adjusted, a few in parallel ways. Others followed different paths. Tropical buffalo were birthed in warmer climes in the south, bison in those colder to the north. More distant in time, much further, they were of common ancestry.

Among the many species far removed from humankind are the elusive, dazzling treasures that I call snow oxen. They survive among cold, icy kingdoms, usually at the tip of a continent or at its loftiest heights. Some are in the deserts, both cold and hot. We know little of their varied lifestyles or near relatives. How much time and what intensity of pressures are required for them to adjust their behavior, alter body forms, modify winter coats, or change mothering abilities? Snow oxen are neither ghost nor hallucination, as the snow leopard was on Peter Matthiessen's spiritual Himalayan journey forty-five years ago, as recounted in the book he titled after the elusive cat. Snow oxen may be relics of the past and fugitives in a modern world, but this need not be the case.

For me, they are metaphors for species of particular landscapes—cold, dry, bleak, high, or at the extremes of latitude—in essence, the future, if we dare imagine conservation of a suite of animals confronting challenge in remote lands. Yet, these species are also real, with ancestral lineages shaped by ice and isolation and whose futures are perhaps cemented by them. Many, however, reveal remarkable abilities to adapt.

"Adaptation" is the critical word here; it's a tricky one with different meanings. It affects how we view which species have rosy futures in a mutable world and what, if anything, they do and we do as the land and the seas grow warmer.

At sixteen thousand feet and halfway to outer space, the air is thin, nearly 50 percent less dense than at sea. The cold is deep. While this could be high up in the periglacial desolation of coastal Alaska, the people I see, along with their unkempt beasts, reveal a different locale. At just 32° north of the equator, a man with almond eyes and dark skin walks with domestic yaks, a mix of piebald, brown, and white streaked. Unlike Inuits, neither whale nor walrus blubber are available to this Tibetan herdsman. For warmth, there is no wood to burn. The gold standard is yak. The shaggy domestics allow semi-nomads to survive the Arctic-like conditions well above the Himalayan foothills. Yak dung is the sole heat source.

Whether for Tibetan, Inuit, or others, animals enable subsistence. If one asked the animals, their first choice would not be to sustain us. We can't ask them, but we can apply the concept of *umvelt* and thereby try to see through their eyes.

What might animals of the wild teach us? Even if we believe nothing or doubt lessons of survival, we do know one thing. Generations upon generations yielded the living forms of today. While internecine conflicts, environmental swings, and even glacial shifts in ice have doomed some species, others continue to survive despite a world of warfare and unthinkable suffering. Knowing the routes caribou travel when winter is severe, why a snow leopard might select marmot over ibex, or whether a herd of horned oxen view people differently than polar bears can all help us assure these animals rosier futures. Perceiving their *umvelt* is one of the tools available to us.

A simple idea underpins my life: maintain what we have and restore what we've lost. A simple but very different idea underlies the field we call science: truth. If we seek truth, two words resonate: doubt and replication. Without skeptical challenge, there can be no certainty. Without replication, we can't be sure. The actual truth differs from the act of its pursuit.

A simple truth underscores biology: evolution. One need not accept it to appreciate nature's beauty. Two-thirds of the bodies of wood frogs in Alaska turn to ice in winter, yet these amphibians survive. A single bowhead whale attains the size of a bison herd numbering fifty. Some mammals lay eggs; some lizards are legless. Bats catch fish. Birds catch bats. Kingfishers fish. Scientists argue that such a range of tailored responses in the natural world is more than pure chance. For nonscientists, it just doesn't matter. But an appreciation of nature's grandeur and our life support systems does matter.

While the habits of many species have not changed across time, ours have. For fun we summit mountain peaks like K2 or Denali. We compete in ultramarathons on different continents. We soar into space and spy on neighbors with drones. Technology offers real visits to the North Pole and virtual ones to our past. This odd mix of individual achievement and modern technology offers hope to improve the lives of humans, yet erodes our curiosity about the natural world. Our growing disconnection from nature continues. Still, none of us are immune from its recoil.

Today we quicken the process of extinctions and paint a bleaker future. Catastrophes occur at the Roof of the World, where temperatures warm two to three times faster than elsewhere. In central Asia's highlands with its forty-five thousand glaciers—the greatest amount of ice outside the two poles—the melting grows serious for the two billion people downstream. From India's Ladakh to Pakistan, lyre-horned antelopes have vanished, and snow leopards are poached. To the north, the loss of sea ice is dooming polar bears and ringed seals. Industrial development, habitat fragmentation, and pure greed are not helping the world's wild animals anywhere. The twin challenges of growing numbers of people and climate modification can generate hopelessness.

Against such numbing prospects, a sense of wonder and optimism persists. It comes from children, from young adults, and from college students. It stems from teachers and a sympathetic public. It gains momentum with the collective voices of people. It derives from the species themselves, well-known and little known, small and large, and from places far beyond the likes of Yellowstone or the Serengeti. It comes from areas few will visit. It emanates from countries like Bhutan and Mongolia, from areas such as Beringia. It comes from western China, nearby Nepal, and from tiny sectors of the globe. The animals themselves are a source of hope and of information. Pandas galvanize, polar bears signal climate alteration. Both do another thing. They beacon conservation.

Given the pace of our alteration of the biosphere, is adaptation possible in a timeliness that can ensure survival? Wild species have, after all, developed unimaginable tactics to cope with environmental challenges. Lemurs with fat tails practice torpor, desert foxes dissipate body heat with giant ears. Subterranean mole rats live as naked of fur as queens, and their workers cooperate in caste systems. Birds use earthen shelters to avoid death. In Africa, Australia, and South America, a particularly adaptable bird carries a generic name—penguin. The point is that some propensity for adaptability exists.

That's the critical issue—the propensity to adapt. How long does it take? Indeed, what does it take to adapt? That's biology, and the questions are not to be swept aside. Conservation, however, requires so much more. We need to know who is listening, and if anyone cares. Once we know, we can ask how to accomplish conservation when adaptations are easy to come by. But, what if they are not easy to come by? For most species, they aren't.

This book is about the odysseys of the metaphorical snow oxen—their lives in extreme environments and their ability to withstand modern challenges. I offer a story based on my own efforts to understand this diffuse group by connecting three themes along common boundaries—seeing through the snow oxens' eyes, seeking truth, and trying to enhance conservation. My narrative is based on thirty-three expeditions, including nineteen in the Arctic (from Alaska to Russia and Greenland to Svalbard), seven in Mongolia, and seven in the Himalayas and Tibetan Plateau (China and Bhutan). These journeys have included learning and working with small mobile squadrons of mostly indigenous people.

I offer three primary vignettes about little known species of the snow ox ilk. The final section ties these together under a conservation umbrella. Of note is that the lives of all these animals remain veiled in secrecy because they are hard to study and because of political impediment. My accounts are glimpses into their cryptic and complex lives.

My pursuit to understand them is far from sanitized. Accessibility issues, weather delays, and government interference always nag. But rewards come—faces of local people who smile when they discover the magic of the wild, the selflessness of teamwork, and the efforts of unspoken heroes bereft of modern amenity. While data are the bottom line for scientists, the conservation line differs. Doing science is not conservation. Donning a human face, inspiring people to care, engaging people who listen, and ultimately persuading decision makers to act is.

It is somewhere along the edges of humanity and remote geographies that I narrate. It is here where wild animals may have a better chance to survive the world's crowded lowlands. Few places though, no matter how distant, are fully safe for wildlife survival. The biological heritage of wild species is assaulted both by people and by modern climate shifts. People living in remote areas, like the rest of us, attempt to improve their lives, behavior which in turn also can be inimical to animals. This book is my journey to unravel surprises suppressed by the metaphorical snow oxen and to determine what they might do better for themselves, and what we might do for them.

A frozen continent lies dark and wilde, beat with perpetual storms of whirlwind and dire hail, which on firm land thaws not, but gathers heap, and ruin seems of ancient pile; all else deep snow and ice.

JOHN MILTON, *Paradise Lost*, bk. 2 (1674)

PART I

The Intersection of Continents—Beringia's Silent Bestiary

If polar bears are the face of climate change, muskoxen are the heart. They're the largest land mammal of either polar realm—Antarctic or Arctic. Neither ox nor maker of musk, their name is a complete misnomer. A nobler moniker is needed.

These regal survivors have ethereal black fur, two-foot horns that unfurl under a massive boss, and thick skirts drooping to the ground. They're an Arctic apparition, a Pleistocene remnant.

Defiant, they stand. They wait, patiently. They have always waited. Summer will come. Winter returns.

They define these turbulent lands, and an uncertain future.

A Pleistocene mother responsive enough to make her baby feel secure was likely to be a mother embedded in a network of supportive social relationships. Without such support, few mothers, and even fewer infants were likely to survive.

SARA BLAFFER HRDY, *The Past, Present, and Future of the Human Family*, 2002

1 Motherless Children in Black and White

The steady din of our small Cessna belies a sense of calm. Below is a wilderness frozen in time. February's windblown snow hides secrets. One is a cluster of dark objects. We bank hard for a closer look. Three ravens scatter. There are fox tracks. Then we see muskoxen.

Adults, each beheaded.

Seven.

Like ghosts, the snowmobile tracks leading from them dissolve. The hunters are gone.

We don't know when the muskoxen were killed or why they are headless. The area is remote. There's no one to ask. The macabre scene plays out in my head. *Who? Why? Did animals escape? If so, who were the lucky ones?*

Surges of volcanic extrusions poke above shallow lakes a month later. Hexagons of ice reflect early morning light.

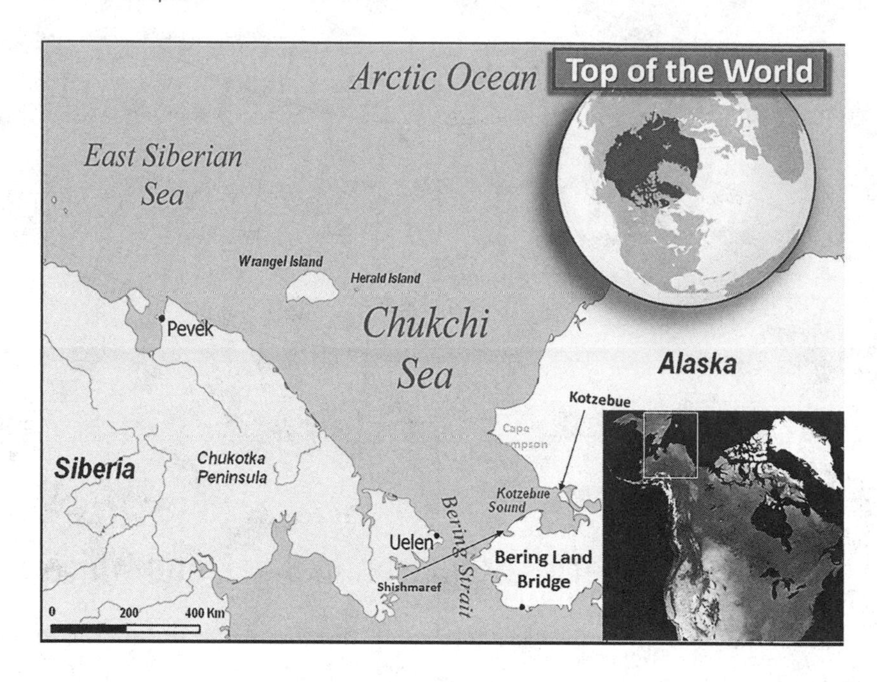

From the air on this March morning in 2010, we're drawn to the dwarf willows below, hoping to spot a flock of ptarmigan warming in sun. Instead, the permafrost is lifeless, patterned. Mountains block the southern horizon. Only the pressure ridges of jagged ice reveal where sea begins and land ends.

This is polar desert, the edge of the continent and the edge of terrestrial life. Our search today is for its largest resident—muskoxen.

For a thousand square miles, we see nothing. Gradually, shadowy rays illuminate beige bodies, and caribou break the wintry monotony. Tracks of a wolverine appear, then vanish. There are a few trails that do not disappear, and these are snowmobile imprints. Some head to Deering, a few to Shishmaref, others to Wales; all are tiny native villages near what was once a land bridge connecting Asia to Alaska. It's now submerged below the ocean but the visible part is protected acreage known as the Bering Land Bridge National Preserve (managed by the National Park Service). The general area is often just called the Bering Land Bridge. None of these hamlets, or Alaska's other 190 remote settlements, are accessible by road, and not even by rail. Each has fewer than three hundred residents, except for Shishmaref. Its metropolis of six hundred is squeezed onto a spit against the Chukchi Sea. A step west is Siberia. To connect to the outside world, only dogsled or snow machine, boat or bush plane are available. Today we are lucky. The weather is good and we continue our aerial search.

Weeks pass. Fifty miles separate me from the carnage of the seven beheadings. I'm on a snowmobile. It's still winter and the snow cover is good, hard. The drifts are tall, the canyons navigable.

My purpose is to locate the living. I want to know whether muskoxen in this part of Alaska have a future. An answer will help me understand a broader puzzle of how warming temperatures and the loss of ice are changing prospects for species reliant on cold. Most of the world knows the plight of polar bears, but the fate of species living primarily offshore will inevitably differ in kind and magnitude from those on land.

I cross more miles of tundra. The rushing air numbs my face. My goggles fog. I warm them on my snowmobile's engine and scrape away ice flecks. Maybe now I'll spot signs of life. My thoughts return to death, to the seven guillotined. *Did herd mates survive the butchery?*

My ears ring from the high pitches of the droning machine. Silhouetted ahead, against a scarp a mile or so away, are black dots. *Basalt?* I lift frozen binoculars for a hurried look. They, too, fog. The dots move. My team of three spew blue fumes in a chase.

My heart pounds. *What are they?*

Closing in we see. They're not adults. Not even juveniles. They're less than a year old—terrified and alone.

Delicate brown eyes bulge, laser focused on our machines. They stand, squeezed together tightly as if hoping to be cradled by mothers who are nowhere around. Steam rises from their overheated bodies. We count them. The number is seven.

Unbelievable—precisely the same as the beheaded corpses fifty miles away. What becomes of motherless children, their psyches, their fates?

High latitudes are special. Every place is, of course. But nowhere are changes in Earth's atmosphere and oceans more dramatic than at the poles. They are literally the refrigeration system for the planet. As the Arctic continues to release more heat to space than it absorbs, the planet's climate will continue to modify, and the Arctic will continue its rapid pace of warming. To dismiss understanding Arctic conditions because they are just too distant, or because most people will never see its wildlife or people, is a mistake. The poles herald our biological world at lower latitude: less ice, more warming, more carbon dioxide, storms of higher intensity, and greater erosion of shorelines.

The challenges are vast and pressing, especially if we care to understand wildlife and people, neither of which can easily be detached from a challenging future. Are the ambient changes at the poles and elsewhere occurring faster than species' capacities to evolve? Can species persist in the advent of radical climate change? What, if any, conservation tactics can effectively be applied? I'm here to explore these issues using animals from the edge—high latitudes and high elevations and a few in between—beginning with muskoxen in a place that once was an immense link connecting two continents. It's called Beringia, a two thousand–mile-wide stretch north to south and an area even longer east to west. It lies at the northern juncture of the Eastern and Western Hemispheres. The terra firma that once connected Asia to North America was literally a bridge. This land enabled human travel between Asia and North America twelve to fourteen thousand years ago; its crossing was regulated by temperature and arbitrated by ice sheets. Water was released or frozen as sea and glacial ice grew or faded across time. With melting, seas grew deeper; as temperatures turned colder, the seas became shallower.

The land bridge opened or submerged. Today it's under water, and most of it has been for more than ten thousand years.

Three years before the decapitations, in spring 2007, I was in a small bush plane north of the Arctic village of Kotzebue, which is an enclave of indigenous Alaskans along the Chukchi Sea. My goal was to find a study area for muskoxen, a species extirpated from Alaska by humans more than a century earlier. They had since been reintroduced.

From the air, I'm circling a group of black dots where the Brooks Range peters out near the frozen ocean. Several additional herds are nearby in Cape Krusenstern National Monument. I've seen enough. A study may be feasible. We touch down on the small airstrip back in "Kotz," and I hustle over to monument headquarters. A government employee greets me.

"Why do you want to fuck around with them muskox?" asks Willie Goodwin Jr., ex-mayor and respected Inupiat elder. I don't understand the question. I expected a grin. There is none.

The prevailing wisdom in the area is that muskoxen compete with and displace caribou. Only a few years earlier, the antlered deer numbered nearly half a million in western Arctic Alaska. In this roadless realm the size of Montana, caribou are sustenance. Neither Willie nor his kin who had settled the region less than five hundred generations ago now had cultural or other ties to muskoxen. Memories are short and people didn't recall the wooly beasts. For Willie and other residents of Arctic Alaska, food security is caribou and sea mammals.

"Muskoxen are just not liked, and can be dangerous. They scare women from berry picking. They chase dogs." Willie had valid points. An additional one rings loudest.

"No one ever bothered to ask us, not when muskoxen were brought back to Alaska, and not now, whether we—the people—had wanted them. It was our land. No one asked."

The Arctic is relevant, and not just because of its climate or global reach. Nor is it just the economic potential of minerals and petroleum products, or even opening its waters for shipping as the sea ice recesses; this last is something akin to the commerce that passes through the Suez or the Panama Canals, though these are all issues of sovereignty. The reason for its relevance to the rest of the world has more to do with biology than dollars.

Among the species of western Alaska that veer into eastern Asia is a bewildering array of fellow mammals. Eleven occur on land, twenty-nine in the sea; one uses land and sea. This quantity of cold-adapted warm bodies exceeds the number of large mammals in Serengeti's diverse tropical brigade. Among the polar land menagerie is the sole species limited to the Asian side, snow sheep. These are graceful but virtually unknown mountain beauties. Restricted to Arctic America are coyotes, black bears, and Dall sheep, although each barely spreads onto Beringia's surface; the bulk of their distributions extend farther south. Inhabiting both the Russian and American sides are moose, muskoxen, and caribou. So do brown bears, lynx, wolves, and wolverines. Polar bears are also in both countries, and live on land and water; as marine mammals, they survive by eating bearded, ribbon, ringed, and spotted seals, all of which also live on both sides of Beringia. The true megabeasts are whales—the bowheads, North Pacific rights, finbacks, blues, and humpbacks. There are herds of walrus and pods of orcas. Human hunters work both land and sea, just as they have for thousands of years.

Beyond a handful of better-known species like bowhead whales, caribou, and polar bears, much of the northern bio-treasures remain off the world's radar. But not so for higher-latitude human residents. Their appetizing fare carries Inupiaq names: *sisuaq* for beluga, *ugruk* for bearded seal, and *inconnnu* for sheefish. Others items remain unnamed, small birds among them. They're not food.

There are other species as well. Nearly half the world's shorebirds migrate to the Arctic. Eighty percent of the world's goose populations visit. Birds like northern wheateaters fly from East Africa;

short-tailed shearwaters arrive from the Tasman Sea, and golden plovers from the tip of South America. More than 275 species are Arctic migrants.

My pressing concern with my muskoxen project is aerial, not avian. It's about helicopters and radio frequencies and GPS collars. To study muskoxen, it's all about logistics—research design, data collection, personnel, and safety. Conundrums are many. How many collars are necessary? How will animals respond when tranquilized? Will herd mates defend or abandon their immobilized friends? If they bolt, can we reunite them? Or, do we heartlessly create orphans?

Might those orphans walk fifty miles?

As it turns out, it wasn't the logistics that haunted me. It was seven orphans.

> Modern humans have followed climate and wildlife since time immemorial. [As] cave paintings and [petro]glyphs of Africa and Eurasia indicate, humans and wildlife have been painfully but inextricably connected from our first cultural moments . . . beginning in the late Quaternary.
>
> STEVEN E. SANDERSON, foreword to *The Better to Eat You With*, 2008

2 Before Now

On an April morning in 2008, the frozen tundra is silent. Rivers and lagoons retain ice two feet thick. The tops of Igichuk Hills rise several thousand feet above the tilting coastal plain, a windswept crown exposing pure rock and desert pavement. Lichen, saxifrage, and moss grace the ridges below. Willows and dwarf alder line river bottoms.

Muskoxen must be somewhere nearby. They're my doorway to the warming world in the domain known as Cape Krusenstern, the national monument mentioned earlier and named for Adam Johann Krusenstern of Russian naval exploits in the late eighteenth century. I'm again traveling by snowmobile. My study has been approved. The snow is spotty; it's deep in some spots, sublimated in others—like ice that's shrunk in the freezer, except here it's just gone.

My team and I inch up steep hills. We navigate Igichuk's abrupt slopes and a few knife-edge cornices. There are no

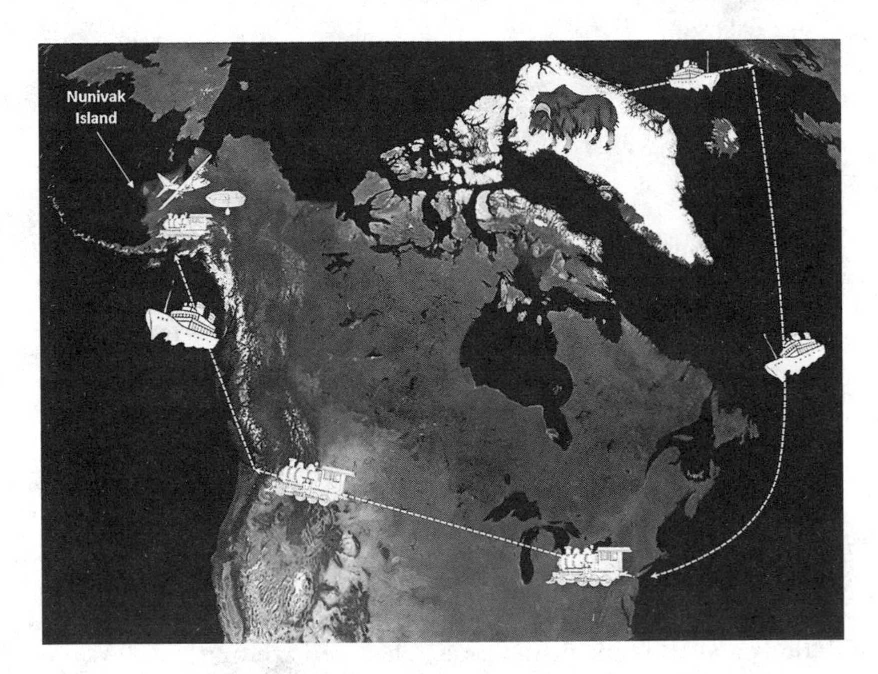

animals. We backtrack but become mired in snow. With shovels and a winch we dig out, only to reach another plateau. Feces come into sight, so we know that herds must have camped here for weeks.

Dark humps break a ridgeline and disappear. We kill engines and approach by foot. A dozen muskoxen are feeding. An item at their feet catches our attention; it's a calf, the umbilicus still attached. The mother sniffs her valued prize. Two emboldened youngsters, a couple of years senior to the newborn, investigate. A lowered head with horns thwarts their approach. The baby isn't more than a couple of hours old.

Ejection from a warm womb into a cold and often snow-laden environment must jolt the system. Today is Earth Day. The wind and cold combine to render the effective temperature 0°F.

Baby muskoxen tolerate temperatures to −20°F. Born with a heavy layer of brown fat, neonates (newborns) are covered with underwool known as qiviut. Finer than cashmere and some eight times warmer than sheep wool, its insulating quality coupled with a calf's high metabolic heat prevents hypothermia. Over generations, the infirm have been weeded out, in spite of careful mothering.

Herd structure, especially the presence of large males, is another feature that improves a baby's survival. One of these males now decides we shouldn't be so near, followed by two females. They stand motionless, their beady eyes staring, with head and horns held steady.

Though being a juvenile is tough enough, being an orphan is worse. As long as the mother is present, a juvenile will more easily survive and perhaps carry on to have their own young. Darwin was pragmatic about parental role: "Maternal love or maternal hatred, though the latter fortunately is most rare, is all the same to the inexorable principle of natural selection."

Two years later, I am wondering about calves, especially the psychological and cultural impacts of maternal deprivation. The decapitations and the orphans dog me. They will continue to, because even by 2017 the mystery remains unsolved.

In about 1720, French fur trappers saw dark, wooly, humped animals in the low-lying barrens of northern Canada. They called these beasts *boeufs musquez*. A full century before, in 1619, early inhabitants of New France—that is, parts of central and northeastern North America—reported a forest bison, a large hairy ox that they termed *le bouef*. Just as Alaskan Inupiat name their edibles, the French, too, used a collective label—beef, steak, or meat. *Le bouef* was bison. *Boeufs musquez* literally were "musk cattle." Though neither ox nor makers of musk, they, too, were likely tasty.

Bison and muskoxen shared the qualities of largeness, hairiness, and darkness. Each also became perilously close to extirpation in the nineteenth and twentieth centuries. The bison case is well known. That of the Alaskan muskoxen is not. The animal's range is broad, and the lands inhabited more onerous. Five countries—Russia, Greenland, Norway, Canada, and the United States—were involved, and boats, trains, and planes were essential to its rescue.

The story of relevant northern mammal extirpation begins four to five hundred years ago in the Atlantic Ocean. The story is the same in the Pacific—just the details differ. The species were whales—notably the blue, fin, humpback, and minke; also, the sperm, gray, and North Pacific right whales were mercilessly hunted. Even during the last hundred fifty years, these were the major catches in the northern Pacific. The narrative is simple and unchanged across five centuries. It's a human story.

People need energy and food. People want profit. Centuries ago, whale blubber and oil generated light and warmth, augmenting candles and heating options for European settlers. Actually, long before that and for thousands of years, Arctic peoples used the products of cetaceans to survive. But with Europeans, a commodity-driven search became increasingly economic. The issue of relevance here, though, is not the whales or their products but their pursuers, the whalers. The munitions that arrived with whalers are what is pertinent to the muskoxen story.

When sailors, and later ethnographers, touched Alaskan shores between the 1820s and 1850s, muskoxen were at low densities. Otto von Kotzebue, sailing under the Russian flag in the 1820s, noted Siberian products had arrived in Alaska, yet guns were unknown. A century would pass until Irving Reed, in 1946, formalized the connections between weapons and wildlife: "It would seem that muskoxen had been . . . on the Arctic plains of northern Alaska up to 1847 when the first whaling ships reached Pt. Barrow. . . . The demand for the hides and the furnishing of the natives with firearms, led to the quick extermination of the two native herbivores—first the much more easily killed and less prolific muskox." The preceding century had seen the introduction of guns with the arrival of whalers, and the resulting effect on some wildlife species was quickly seen.

By about 1890, muskoxen were gone, but a move was soon underway for a reintroduction. In 1930, Territorial Governor Scott Bone sought money from Congress "to aid in conserving a species threatened with extinction" and "for contemplated experiments in reestablishing the muskox as a native animal in Alaska." With his request approved and $40,000 available to the U.S. Biological Survey, the next challenge was to identify a source population and sort out methods of capture.

The helicopters, veterinarians, and tranquilizing drugs so common now were unavailable in the 1930s. How did one capture muskoxen? Norwegian sailors provided the answer.

"There is much violence in a flock of muskoxen," wrote a Danish expedition leader to Greenland, after his men shot adult bulls were so that they could rope calves and yearlings and not get crushed. In today's parlance, the capture would be called shockingly brutal, a bloodbath, a slaughter. Mass killing of herd members is the same way elephant overpopulation was dealt with in Zimbabwe and South Africa; they were shot in groups to remove each and every traumatized individual.

Alternatives were not on the table, at least not for the tightly bonded herds of living muskoxen in the 1930s. Adults were dangerous. They still are.

The only way to capture calves was to kill all of the adults in a herd. Reports from earlier efforts to capture calves include Gustaf Kolthoff's 1900 *Til Spetsbergen och Nordöstra Gråonland år oirts*:

> We went around the animals and approached them so that they found themselves between us and the sea. As soon as they became aware of us, they stood immediately in a protective ring, and calves took their place among the elderly. So the magnificent animals lowered their heads threatening to turn against us, and when we reached them at good shooting range, I asked Kjell to shoot first. He hit a bull in the forehead, who fell instantly. At that moment I shot one, and a third went to Levin's shot. Now the other animals took to flight, but in that instant a big bull got a bullet from my second gun, with the result that he drove his horns into the ground, did a somersault and

> tumbled dead near the embankment down to the beach. Only two cows escaped in the company of the calves. I now switched out the double paradox rifle with the mauserstudsaren and thus brought down one cow, while Kjell shot down the other. The calves then stopped and became encircled by us....

Also, a report from a 1904 Danish report provides a strong dose of reality:

> Divided into two parties we went down each side of the flock. The calf walking a little above the others together with a cow it had sucked a little before... sprang up to the others, that instantly formed a circle around it.... All the animals turned their heads outward.... Snuffling and snorting, their noses turned toward the earth, the animals were ready to receive the enemy.... By the time the first two animals fell, it was strange to see a trembling go through the herd, the animals pressed closer to the calf, by the next 4 shots, 3 animals fell and a bull went away shot in the lungs.... The two cows now left pressed the calf between them; then one fell bleeding, the other protected the calf.... We now tried to chase the [remaining] cow on to the beach; she had sought... cover beside the killed bull and the little calf pressed closely to her. The cow walked a little but then turned toward us, fell and died. The calf took up its position at the hind legs of the cow... but it slipped away again [at our approach], still it ran only around the cow and it was not long before it was tied.

Accounts about killing and survival (not for reintroduction) from northern Canada differ only in small details. Another writer, W. T. Hornaday, described the vulnerability of young muskoxen:

> In March 1901 Captain HH Bodfish sent out from the Whaling Steamer *Beluga*, then wintering in the Arctic Ocean north of

> Great Bear Lake, a party of whalers and Eskimo hunters for the purpose of capturing live Musk-Ox. About 30 miles from the coast the party encountered a herd containing four calves, all four of which were finally captured. Unfortunately, however, two of the calves were soon killed by the savage sled-dogs. The two remaining animals were harnessed to sleds, and driven back to the ship. In a short time a third calf was killed by the dogs leaving only one specimen a female about one year old.

In Iceland, where large mammals never roamed, the dream of an enriched fauna was almost realized in 1929 when Reykjavik's Central Austurvollur Square brimmed with six calves. Thirty-four adult muskoxen in Greenland had paid the price. Worse still, they died for naught because none of the captured youngsters reached adulthood, and all perished from local diseases. Two years later, five more calves were brought in from Norway. Again, none survived.

Conservation was a very bloody business. It still can be. For muskoxen, calves are killed during the process of capture for reintroduction. Others die of disease. Some stand with deep eyes glazed, staring mournfully. Others remain next to their dead mothers. Those unable to run are kept in place by dogs. Inevitably, some are captured. The progeny of some surviving orphans that began a journey from the shores of the eastern Greenland in 1930 ultimately became the foundation of all Alaskan muskoxen today.

The babies that endured the Greenland slaughters were first taken to Norway. From there, nineteen females and fifteen males were crated and sent by steamer to New York City. In September 1930, the director of the Bronx Zoo was asked by border authorities for advice about muskoxen husbandry before the eventual transport of the animals to Alaska. The wayfarers were quarantined in New Jersey for thirty-three days.

Subsequently, they were sent more than twenty-five hundred miles by rail to Seattle and spent seven days on another ship, which then docked in the town of Seward, Alaska, far from the Seward Peninsula. The muskoxen then took yet another rail ride, this one north to Fairbanks. After a couple of years at Alaska's Agricultural College, twenty-seven were boated down the Yukon and Tanana Rivers to the Kuskokwim Delta. The final destination was Nunivak, a seventeen hundred–square mile volcanic island out in the Bering Sea, twenty miles from the mainland. From these original transplants, the population grew to more than 650 animals by 1970.

The Nunivak population became the source for subsequent reintroductions elsewhere. One such reintroduction was of sixty-four animals taken during the 1970s to the Arctic National Wildlife Refuge in northeastern Alaska.

Another seventy-two went to the Seward Peninsula (in 1970 and 1981), while another two sets traveled to the Cape Thompson region north of Kotzebue (in 1970 and 1977). So while the introduction of guns by whalers in Russian Alaska before statehood brought the demise of the area's muskoxen in short order, the restitution effort involved many years and crossed expansive geographies. As a result, some muskoxen traveled from Greenland to Norway and across the Atlantic. From there, they were shipped across the United States by railroad and then to Beringia through Canadian and Alaskan waters. The little known muskoxen reintroduction effort remains a remarkably silent victory.

Like species everywhere, there is variation in muskoxen numbers and the rate at which populations change. Some proliferate, some don't. In the Bering Land Bridge, the population increased at about 15 percent a year for thirty years. To the north of Kotzebue, from Cape Krusenstern to areas farther north, the population appeared to stagnate for a decade after arrival. Both areas are managed by the National Park Service.

What was it about muskoxen living in the Bering Land Bridge that allowed their numbers to increase but prevented a similar increase in the Cape Krusenstern realm? Each population is at the continental edge, adjacent to the Chukchi Sea.

When I initially considered a project on muskoxen in 2006, I wondered about the possibility of a comparative study in two remote areas. Later, I enlisted the help of Layne Adams, an Alaskan biologist with an eagle eye, a steady aim, and experience shooting some fifteen hundred caribou with tranquilizing darts. The park service helped secure funding, and together with additional support from the U.S. Geological Survey, a project emerged in 2008.

The differences in population growth between Cape Krusenstern and the Bering Land Bridge could stem from food limitation, alteration of pregnancy rates, or variation in adult and juvenile survival. Remote sensing will help reveal seasonal patterns of vegetation growth and what roles temperature, snowfall, or rain have played. At the ground level, I'll assess pregnancy and stress by monitoring hormones gathered from feces and use these data when contrasting population differences. Although I won't initially know whether the pooper was male or female, later DNA analyses will help distinguish the sexes. This won't be rocket science, but once we know which traits are associated with increasing or declining populations, we can begin a search for causes.

I'll photograph juvenile muskoxen using a process called photogrammetry. Head and profile shots at known distances from the animals will enable estimates of face sizes. I can then ask whether individuals from different areas vary in head dimensions and if they grow at different rates. Such metrics will help interpret nutritional status, because muskoxen on poor diets grow more slowly. No surprise there, as the same is true for humans. Then I may detect signals that climate affects vegetation by evaluating the rates at which young muskoxen grow.

Beringia spans three countries, including the northeastern sector of Siberia, Alaska's Arctic with a long extension of the Aleutian Island chain stretching geologically close to Kamchatka, and the far northern reaches of Canada's Yukon Territory. Due to its dryness and associated habitat features, Beringia held a very different Pleistocene fauna from the species we know today. Its uniqueness was, and still is, unparalleled on earth. Among the denizens were camels, giant bison, wooly mammoths, and even rhinos and yaks. There were caribou, elk, and muskoxen, as well as saiga antelopes, and several wild ass species. These prey attracted carnivores—Asiatic lions, short-faced bears, hyenas, brown bears, wolves, and wolverines. Species like mammoths and giant bison eventually died out completely, as did wooly rhinos and short-faced bears.

Others survive today, but no longer in Beringia. Saigas and wild camels live in central Asia, and wild yaks remain only on the Tibetan Plateau. Caribou are circumpolar, and elk—also called red deer—and wild bison inhabit regions south of the Arctic and in Europe, Asia, and other parts of North America.

Today, the distance between Chukotka, on the Russian side, and Alaska's Seward Peninsula is less than sixty miles; the countries are separated by water at a depth of only one hundred meters. Many lakes are far deeper. Nowhere in northern high latitudes did a continental connection like Beringia exist, but due to the land bridge, numerous Alaskan plants and animals are Asian in affinity.

Indeed, Alaska was peninsular Asia at times in its glacial history. While the land bridge was traversable, species were prevented from moving farther east into western Canada or farther south in Alaska because two massive ice sheets adjoined; they extended thousands of miles and were several thousands of feet deep. Alaska was truly Asian, a fact noted early by the Russian oligarchy in Saint Petersburg when they sent a sailor named Vitus Bering to explore and claim the eastern Pacific lands as Russian. That was in the 1740s, and Alaska remained under Russia's rule until purchased by the United States in 1867.

As in all places, Alaska's populace varies in attitudes and in survival strategies. Cold weather often brings forth common challenges and a spirit to help. Group living comprises an essential ingredient in which cooperation molds success, such as lodging together or sharing food, including a harvest (like the hunting of bowhead whales). The success of my investigations of muskoxen is also likely to depend on local support, and this possibility raises the specter of project transparency to new levels. I'm anxious to communicate, but this gets tricky because I'm not tribal Alaskan.

Areas beyond the state's borders are referred to as "Outside," a term that connotes the lower United States, regions where non-Alaskans make their livings and, at times, project their differing views northward. As an outsider, I'll need to explain the science and logic behind my muskoxen project to people on the ground in Arctic villages and in places like Nome and Anchorage; I'll also have to explain it to a broader American public.

My first presentation will be in Kotzebue. It's the National Park Service's headquarters for the western Arctic. People here are not glued to computers, and in 2008, the onslaught of social media is still a few years off. Advertisements for my impending talk are displayed at the public library, the post office, and the community center. Postings also grace the entries of Koztebue's two general stores, places where milk and cheese, salmon sausage, and even avocados are found. So are wolf pelts and carved whalebones.

Late April is still cold, but spring is everywhere. By afternoon, water pools in the streets. People may still be buried in cabins, but the local radio station (KOTZ) beams its signal with ten thousand watts; it's the single media outlet embraced by most. I agree to a live interview. The first caller asks directly about muskoxen: "When can we start shooting them?" The question makes sense. In an area where Saltine crackers fetch $9 a box, avocados cost $4 each, and peanut butter containers are $6, I'd also rely on sheefish, caribou, and *boeufs musquez*.

With an average annual per capita income of about $25,000, Kotzebue, as judged from the outside, is poor. Full-time employment is low and college degrees a rarity. Ten percent of the homes house three or more generations. But from the inside, the view differs.

In summer, people go to country camps to fish, pick berries, and shoot. The roads in town are paved. There is the $42 million Maniilaq Health Center, as well as evident community spirit and pride. The Kobuk 440 dogsled race and ice fishing take place in winter. There are snowmobile races, the machines known locally as "snowgos." Even without races (snowmo)bilers scream up and down streets and on the iced inlet, often until 1 A.M., when the clatter fades along with daylight. There are outdoor footraces, and indoor dances, basketball, and volleyball. For cuisine, Kotz has pizza and burgers, Korean, Thai, and Japanese fare. Lattes are $6, and sushi plates three times that. Gas can reach $7 a gallon. There are downsides besides the prices. Like everywhere, domestic violence, petty crime, and an occasional murder occur.

Well-known people pass through, including the likes of the once starry couple, Angelina Jolie and Brad Pitt. The most important visitor was Barack Obama, the only sitting American president to see the Arctic and grasp climate change issues while getting a quick look at the lifestyles of native Alaskans.

With increasing day length and warmth, long-dormant objects poke through the melting snow. Dilapidated trucks, boats, and shipping containers emerge, as do antlers piled on roofs. Winter's trash appears. Ravens and dogs delight, especially enjoying the fish remains tossed back onto the sea ice. Snow buntings have already passed through, as they do every year. Gulls have arrived, and geese are overhead.

My trepidation grows as my talk nears. I need to sell the idea of why the unpopular muskoxen should be studied and why someone like me is appropriate and, also, to communicate a broader message of relevant science. My unease is heightened by local confusion as to why someone from the outside will be researching "Alaskan" animals.

The National Park Service will sponsor my presentation, although its resource management creates much dissonance in Alaska. It often matters little locally that Alaska's national parks are public properties, part of an American legacy, or that some 280 million people visit America's 406 national park sites spread across fifty states and territories every year. The park units in northwestern Alaska—the Noatak, Kobuk Valley, Krusenstern, and Bering Land Bridge—receive just a handful of these tourists. Adjacent Alaskan communities reap few rewards, economic or otherwise. These lands of park designation were established only in 1980, which means that not all that long ago they were appropriated from the state of Alaska, from the Alaskan residents, from people whose ancestors have been living here for twelve thousand years. The land transfers remain a source of rancor, and the tension often surfaces yet, especially in areas beyond Anchorage.

My thoughts turn toward adaptation, when thinking about my upcoming talk. I realize I'll have to convey its importance in how animals, including muskoxen, adjust to changing environmental conditions, including climate change.

In the 1970s, two geophysicists working for Petróleos Mexicanos, the state-owned petroleum conglomerate, came across evidence of an abrupt and cataclysmic alteration of the environment. They were searching for energy sources on the northern Yucatan Peninsula but stumbled across something more valuable—the cause of Earth's greatest mass extinction. It was the Chicxulub Crater, the site where a meteorite six miles across smashed into the planet's surface 65 million years earlier. All kids and adults know the consequences of that impact: dinosaur extinction. In short, sauropods did not adapt.

Likewise, everyone knows that when it comes to less sudden changes, environments still challenge species in different ways. Sand dunes in the Sahara and Death Valley reach temperatures in excess of 160°F. Species that survive there are not the same as those of Beringia. Polar bears rely on floating ice. Giraffes live in some deserts. It's safe to say that bears do not occur in sand dunes. It's also safe to say giraffes do not kneel to drink where lions lurk. Actually, a few individuals might have, but they probably did not have long lives.

Where predators are extinct, giraffes do sometimes recline. Here's the point: some species can show flexibility in what they do and where, and in that way may be able to adapt more easily.

Back to sand dunes. Massive dunes also occur in Alaska, and these hills heat to more than 100°F, akin to those in true deserts. Kobuk Valley National Park encompasses twenty-seven hundred square miles above the Arctic Circle, east of the Chukchi Sea, and includes several areas of dunes. Bears, both black and grizzly, occur in Kobuk, though rarely in the thirty square miles of pure dunes; the bulk of Kobuk's habitat is a mix of boreal and tundra, where both species hibernate. Much farther south, in places like Florida and Mexico, black bears do not regularly hibernate. Even at Lake Tahoe in California and Nevada, some are active all year. Grizzlies also once occurred as far south as northern Mexico; though now extinct there, the cause was not oversleeping. Across a species' range—bears, in this case—food availability is regulated by seasons, and its influence is reflected in the adaptive capacities of individual bears. Where food is unavailable, hibernation rules; where available, bears are more likely to roam in winter than live off their fat (catabolize) in a den. Variations on the hibernation theme occur, and these affect capacities to adapt to different environs.

Animal adaptation in the face of change—how quickly, where, and when—is the issue of immediate interest, regardless of whether the shift in environment is caused by humans or not. At a scale smaller than meteors and dinosaurs, consider how increasing urbanization affects animal lives. Not ours but, instead, a small bird with a black cap, a black bib, and a home in northern mixed forests of Canada. The species is the black-capped chickadee.

Like humans, chickadees are physiologically sensitive to high traffic volumes but not to traffic jams at rush hour. Their sensitivity stems from the heightened auditory clamor, because noise affects detection of song, and this is a serious problem for males whose melodies are broadcast to attract females. Scientists from the University of Alberta seized this opportunity to explore if or how chickadees handled their urbanizing and increasingly noisy Edmonton environment. While obviously not quite sauropod in size, and with a bird brain, the question of adaptive capacity in the short term is relevant. It sheds light on how species respond when environments are altered, irrespective of source—whether meteorite-induced atmospheric change, road noise, or climate.

Chickadee responses are noteworthy. When traffic levels are high, male songs become higher pitched; when noise abates, songs are lower in frequency. This is comparable to people trying to talk at a loud bar who amplify their voices when it's hard to hear, and soften them when it quiets down.

That chickadees figured out what to do when conditions changed rapidly is useful when considering the past, including the inability of dinosaurs to survive or the current situation of muskoxen. Plasticity in behavior occurs for many reasons, including adjustment to local conditions. It is these behavior variations expressed by individuals that become the true grist for natural selection, the mechanism by which ecological, or in this case behavioral, processes drive evolutionary change. The pace at which change occurs is a fundamental issue both for conservation and for understanding species.

The significance of variations is an idea that was influential in modern biology and scripted by Charles Darwin in 1859 in his *Origin of the Species*: "This preservation of favourable variations and the rejection of injurious variations, I call Natural Selection. Variations neither useful nor injurious would not be affected by natural selection and would be left a fluctuating element."

Alfred Russel Wallace was Darwin's lesser-known coauthor of the scintillatingly titled 1858 piece submitted to the Linnaean Society, "On the Tendency of Species to Form Varieties; and on the Perpetuation of Varieties and Species by Natural Means of Selection." Wallace later explained his realization of the importance of individuals: "Why do some die and some live? And the answer was clearly, that on the whole the best fitted live. From the effects of disease the most healthy escaped; from enemies, the strongest, the swiftest, or the most cunning; from famine, the best hunters or those with the best digestion; and so on. Then it suddenly flashed upon me that this self-acting process would necessarily improve the race, because in every generation the inferior would inevitably be killed off and the superior would remain—that is, the fittest would survive."

Whereas neither dinosaurs nor wooly mammoths survive today, muskoxen do. Their response to today's rapidly warming world and other challenges remains uncertain. Darwin and Wallace were interested in patterns across time and space. My focus is here and now.

I'm the outsider at my own talk, which is tonight. The small audience of twenty is already confused. Why, after all, would someone from the University of Montana who doubles as a biologist with the New York City–based Wildlife Conservation Society (WCS) study Alaskan muskoxen?

I begin with images of a warming Arctic and touch briefly on my past work with Alaskan moose and caribou, mostly for context. Then I pose the question of why different populations grow at different rates than one another. Is it due to predation, weather, food, or something else? I describe how remote sensing will help us identify vegetation patterns and snow and ice distribution. I also explain how the noninvasive nature of photogrammetry can inform our understanding of a possible relationship between muskoxen head sizes and nutrition. So far, so good.

I then reach a delicate point. I decide to put forth the plan to immobilize more than thirty animals in each of two study sites (the Bering Land Bridge and the area between Cape Krusenstern and Cape Thompson). Layne is away in Anchorage, and so I'm on my own when it comes to answering any questions. These begin with Seth Kantner, a well-known Alaskan author. With his voice raised and slamming his hand on the table, he heatedly criticizes the study.

"Too many—too many animals to be drugged!"

I pause, taking a deep breath.

"I'm not sure what you are saying."

"That's too many, it's not right."

I thought about the direction this might go, and replied, "How many animals do you think would be appropriate?"

"You're the biologist, you tell me."

As I try to explain why sample sizes matter, Jim Dau, the experienced local biologist working for the state's Department of Fish and Game, then proceeds to neuter whatever minuscule progress I've made. Jim reminds the audience of his vast experience and says that he and his agency know everything needed for muskoxen management. The move was brilliant, if his intent was to cut my team off at our knees. It worked.

Audience doubt trumped prospective science. Perception mattered, not facts. To be relevant, I'd need to do better.

> Muskoxen . . . cannot go farther north and they live right up to the edge of glaciers. They are literally living on the edge of terrestrial existence.
>
> DAVID GRAY, *The Muskoxen of Polar Bear Pass*, 1987

3

Beyond Arctic Wind

Winter weather controls my life. For muskoxen, the day's weather doesn't matter. They have fur. In storms, they lie down or stand. Body plan dictates survival, just as survival has engineered their bodies. Form and function are ingredients of natural selection.

Muskoxen are energy savers. In winter, they move little. Walking burns calories, and running even more. Staying put equals energy savings. In addition, their stubby legs are not good for navigating deep snow, and repeatedly sinking and hauling their bodies from the snow is exhausting. When snow is hard-packed and ice-crusted, the muskoxen dig craters into it, either by pawing or, when that becomes too difficult, by dropping their heavy heads onto the surface, to get to the lichens and mosses below. The food found there is converted to energy and, as long as they can feed in that one spot, calories are saved by not having to dig craters again.

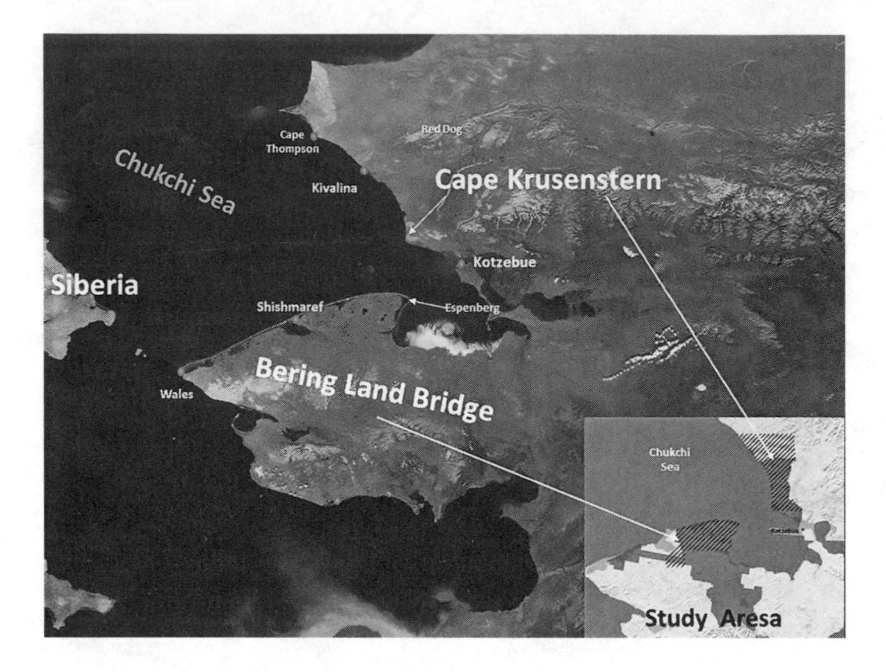

Cold weather tactics differ by species. Ptarmigans, foxes, and hares retain heat in snow burrows. Squirrels do so with nests. Bears hibernate.

Muskoxen are aboveground versions of ursine fatties. They have no snow holes. They seek no ravines. Instead, they nestle down together. Furred skirts protect legs, but when their adipose supplies run out, they die.

Wind and visibility will affect my comfort level and my ability to find muskoxen. With the goal of understanding if and how climate affects the demography of the two muskoxen populations under study, logistics nag—particularly the navigational abilities of small aircraft and helicopters. This means wind and visibility matter. I will need aerial support to dart animals so we can affix radio collars. But for the fourth straight day in one of the early field seasons, Layne and I are grounded, waiting for a storm to abate.

Pilots are not fond of flying when sky and earth combine into muffled Arctic grayness. That recipe grows more serious without sun and when having to spin wildly to dart animals. The difference between knowing up from down defines life and death. Today's hazardous conditions dash all hope of using the Hughes 500 helicopter.

Less worrisome are ambient temperatures, at least as long as they're above −15°F or so. The need for safety and efficiency, however, do not subside. Darts become brittle and shatter. Narcotics like the opioid Carfentanil—our tranquilizing drug—can freeze inside the dart. Even when we are successful, tranquilized animals quickly lose heat when it's this cold.

Warm and working fingers are always good when handling immobile animals. Gloves help with the warmth but hinder dexterity, which I need in order to affix the tracking collar, fasten nuts and bolts, and remove thick hair around an animal's neck. I also need agility to retract the muskox's lips to examine teeth, and, just as critical, to feel for the thumb-sized tail and the anus. Once my fingers reach balmy bowels, they will warm as I hunt for chip-like nuggets of poop. Such pellets are highly prized because they'll be from known collared animals whose pregnancy status can be determined later from blood sera, which Layne has drawn.

Layne is also the darter for our operation. Rick Swisher is the aviator, an archetypical pilot who is old and bold. And, he's learned lessons by surviving a crash or two. As we wait out the storm, Rick paces because he's needed for a wolf capture five hundred miles away. Layne's already booked a flight back to Anchorage. Thus, with only a narrow window of time scheduled for our captures, we hope for better weather. Still, a certain Zen-ness is the ticket for survival—and for research.

Our plan is absurdly simple. Find a herd. Drop me on permafrost away from muskoxen so Rick and Layne can maneuver the chopper more easily. I'll remove Layne's door to enhance his darting prowess.

The experience will be new for me. Other than a satellite phone with limited battery life in extreme cold, a two-way radio to call Rick, and a GPS unit to report my location, there is no backup. Well, that's not quite true: if there's a problem, I'm to contact the Denali emergency response center 450 miles away. I'm not sure what they could do, at that distance, but having a number to call makes my mom feel better.

The next morning dawns a brilliant orange. The southern horizon is clear. The National Oceanic and Atmospheric Administration weather report is for a tepid sun. We fly north up the Noatak River, which is white, winding, and ice covered. A moose mom and her calf kick up snow. Fox and hare tracks abound. The remnant spruces at the northern edge of their range disappear, the farther north we fly. To the northeast is the Brooks Range.

Angry gray clouds to the west tell a different story there. The sea ice has been torn open from the strong winds of the past few days. The fractured ice has created huge leads—that is, the channels of water splicing through the sea ice. The collision between the warmer water and the frigid air above produces a moisture bank of fog. Long leads mean seals; seals mean food, and this provides perfect conditions for polar bears.

We fly over undulating hills and valleys until we spot twenty to thirty muskoxen. The helicopter then turns sharply and lands. I remove the door as Rick bellows:

"It costs several thousand dollars. Hold it like a baby and don't let it get fucking scratched."

I dip under the spinning rotor, door under arm, and walk deliberately in front of the chopper. Safety first. Rick smiles.

Like a moth, the Hughes disappears as it is swallowed by the bowels of the vast Arctic. The tundra grows silent.

I have no sleeping bag, no shovel, no saw for ice. And no gun. My inflatable rubberized bunny boots, truly vintage World War II leftovers and rated to −40°F, help my toes. I wear one face mask but carry another. Wearing two pairs of gloves and shoving my hands inside a windproof shell keep my fingers alive. As I wait for the helicopter's return, I exercise and look for tracks of hare or wolf. My pleasure is inversely related to the strength of the breeze.

From my hillside vantage, I see coal-colored bodies two miles off. The muskoxen coalesce into a defensive group, a sure sign that they've detected the boys in metal pulsating above. Chaos reigns. Twelve flee in one direction, ten or so in another. The herd has split.

Conquer and divide is good in warfare and for police doing crowd control. It's not so good for muskoxen. Despite the scientific literature describing the animals as prudent expenders of energy, these dark Volkswagens don't read. They pound off calories when they flee through deep snow. I worry about them suffering leg damage and broken bones. But thumping above them is the reverberation of hammering air.

Layne leans out the doorless opening and takes aim. The helicopter drops over a hill. I see nothing. The radio goes silent.

The most prominent large mammals of the world's current polar landscapes are caribou and muskoxen. Beringia lost its wild horses and assess in the late Pleistocene as vegetation changed due to climate, although either they or close relatives do persist elsewhere in Asia. Gone everywhere are giant bison, short-faced bears, and mammoths. The causes of those extinctions are not universally clear and, indeed, climate is just one possibility. Predation by people is yet another, especially when species were fearless because they had never been exposed to humans before Asian hunters crossed into North America. Other possibilities exist as well.

Drought might have reduced food availability, or the greening of vegetation may not have occurred at the same time the requisite protein was needed to support lactation. Maybe predation by nonhuman animals was relevant.

Whatever the cause or causes, the bottom line is that it's less possible to project a species' future without understanding both weather and interactions with other species, and the changes in intensity of each across differing environs. That comprehension includes recognizing the nuances of survival at the edges of the species' range. This is what I'm trying to understand here for muskoxen.

Caribou and muskoxen in Alaska both live in groups despite vivid differences in body form and predator defense. Caribou coalesce in large herds and flee when in danger. Muskoxen usually face it down.

Life in groups can be a part-time enterprise for some animals. For them, some time is spent alone. For others, the group is life itself. If individuals of group-dependent species like muskoxen are placed in solitary confinement, they become stressed. Some die. Muskox society is defined by groups, whether on Siberian islands, on the Asian mainland, or in Greenland, Canada, Alaska, or Norway. Females live with females year round. Most are attended by one or more males. Males are more variable. Sometimes individuals are solitary, but more frequently they associate with just a few others.

Food assuredly dictates total numbers, and therefore group size, but other forces operate as well, especially predation. More aptly, it's the threat of attack that affects decisions about security. Darwin knew this, writing in his 1871 book *The Descent of Man and Selection in Relation to Sex*: "With those animals which were benefited by living in close association, the individuals which took the greatest pleasure in society would best escape various dangers; whilst those that cared least for their comrades and lived solitary would perish in greater numbers."

Caribou offer a fine test of Darwin's proposition, particularly regarding group size and solitary living. In Alaska, grizzly bears and wolves are key predators. Caribou herds are impressive, with some numbering in the thousands. Yet, on the islands of Greenland and Svalbard, just 10° south of the North Pole, grizzly bears have never occurred, and wolves have only periodically lived in northern Greenland. They did not make it south, and never to Svalbard. Having no enemies for a long time, caribou on these islands might be expected to relax their defenses and perhaps fail to register danger, similar to some naive birds of uninhabited islands—much as Darwin reported from the Galápagos in 1835: "A gun here is almost superfluous; for with the muzzle I pushed a hawk off the breach of a tree. One day, whilst lying down, a mocking-thrush alighted on the edge of a pitcher... which I held in my hand, and began very quietly to sip the water."

While crossing Greenland as part of the Thule Expedition in 1912, Knud Rasmussen reflected on the lack of fear in wildlife: "We encountered a hare so amazingly tame that we were tempted actually to essay his capture with our bare hands. Soon afterward we spied a lonely caribou who at once was all curiosity and came running toward us to investigate these strange visitors. The confidence of the game showed well enough how little disturbed the region had been."

This, too, was evidently the case for four Pomor sailors from the area now known as northwestern Russia who, in the 1600s, became stranded on Svalbard when their supply ship drifted away one windy evening. Weaponless, they had a real problem. How would they obtain meat?

They solved their dilemma by crafting arrows from driftwood and rusted nails. Over the course of four years, they shot two hundred fifty wild reindeer. (Reindeer and caribou are the same species; their names just differ by geography.) They've lived on Svalbard for thousands of years in the absence of humans or other predators. Given the crudeness of the Pomor weaponry, reindeer must not have been very wary, or the men expert stalkers.

To try to determine whether the lack of fear and reduction in group size on islands were possible responses to the lack of predators, I experimented with using a speaker and amplifier to broadcast wolf howls. For my control, I broadcast the sound of running water. Caribou in Alaska became more than three times as vigilant at the sound of howling wolves—lifting their heads, turning their ears, and failing to feed—than their remote peers on the wolf-free islands. When the sound of running water was broadcast, the response of the caribou did not differ by locale.

One more behavioral observation: during these broadcast-sound experiments, Alaskan caribou clumped by running to other caribou—"comrades" in Darwin's words—some nine times more than did their naive island brethren. Whichever lens one looks through, the threat of predation shaped how this northern species organizes its spacing and behavior.

These sorts of contrasts in wildlife behavior on islands and mainland areas are illustrative of how species or populations can change with environmental conditions. In this case, the absence of predation affected behavior and social structure. On the wolf- and grizzly bear–free Arctic islands, not only are caribou group sizes small, but females are also frequently solitary. Some 25 percent of the caribou I observed on Greenland and Svalbard were either alone or with one other. Such small groups never occurred in Alaska.

Is the Arctic's other large social mammal of land—muskoxen—more like caribou or elephants? Both Asian and African elephants live in fusion-fission societies, where individuals periodically come together and leave one another. Caribou do, too, but a key difference is that elephants, unlike caribou, have strong social bonds and live in extended family groups. They protect each other and sometimes attack predators. For muskoxen, little is known about social recognition, bonds, or the effectiveness of their defensive formations when interacting with capable predators. Whereas it is well known that elephant recognition of herd mates is strong, far less is known about wild muskoxen.

Size itself confers an advantage, and elephants are often considered immune to predation; this is untrue for muskoxen, which are less than one-tenth the size of elephants. Despite their massive dimensions, however, pachyderms are very touchy when concentrated around water and displace dangerous predators like lions, as well as jackals and spotted hyenas. Herds aggressively promote juvenile survival. Older matriarchs at Kenya's Amboseli National Park coordinate family groups in charges against lions at the sound of roars and approach those made by male lions with a greater intensity.

Scottish biologist Graeme Shannon, reflecting on the elephants' actions, has said: "We believed that this behavior was aimed at flushing the lions and therefore removing the threat of ambush attack. Indeed, a group of charging elephants is formidable and I doubt any lion would attempt to attack the young, especially as they were almost always kept safe at the centre of the group."

While calves are usually well buffered from lions, a few exceptional cases demonstrate that, when the cats are bold, they can penetrate elephant defense systems. These cases of lions preying on elephants have been restricted to Botswana's Okavango Delta and are still not clearly understood, but they may have relevance to grizzly killings of muskoxen that I'll discuss later. The lions from Chobe National Park there became elephant specialists in 1985. Over the course of four years, they killed at least eleven, seventeen, nineteen, and twenty-seven elephants (respectively), all but one of whom were females and young. Why the restricted predatory behavior of lions has not regularly expanded throughout Africa isn't clear. Maybe lion prides elsewhere have, in fact, tried ambushing elephants, but researchers missed the action. Or perhaps the lions themselves were killed in attacks and so such deadly tactics not repeated by others. African elephant social systems and associated defense, coupled with their large size, have generally rendered them reasonably immune to predation by animals other than humans. As alluded earlier, while much is known about herd structure among elephants, the same cannot be said for muskoxen.

While it's still uncertain what the major cause of muskoxen mortality is now—food limits, disease, or predation—there has long been speculation. The effectiveness of the unusual collateral (herd) defense of muskoxen is described in Harry Whitney's 1930 observation from Greenland, where wolves periodically survive in the north. "The noblest of the big game arctic animals is the musk-ox, and they are putting up a better fight against the Arctic wolf and polar bear than some of the other species. They have a method of defense all their own,—forming a circle and keeping the young within it when attacked by wolves or when the hunter gets too close. The old ones will charge with terrific force and speed, and then, without taking their eyes off the object of attack, they will back into their circle again with the young safe inside it."

Intuition suggests that such defense must be effective. How, otherwise, would such behavior come about? A gaggle of heads and horns facing down and outward, swirling and spinning—daggers to be thrust—is certainly one of the ways that human hunters were prevented from capturing juvenile muskoxen. As with elephant calves turned away from the outside, young muskoxen are pressed toward the center of the group, occasionally taking their own glances at whatever has prompted the huddle-like defense.

As effective as a standing defense may be, it did not arise on its own. Body plans develop from environmental pressures and other forces; the result is a compromise among what's effective, possible, and constrained. As noted for Greenlandic and Svalbardian caribou, behavior is different when there is no predation.

Other evolutionary forces prevail with isolation. Island species can experience dwarfism, a reduction in body size that has occurred not only to insular forms like dwarf hippos (on Cyprus) but also to a member of our genus *Homo*; on the Indonesian island of Flores, *Homo floresiensis*—the popularly named hobbits—stood three and a half feet tall. Svalbard's reindeer have also become dwarfed. Although living below glaciers—where we might otherwise expect animals to be larger in size because cold-adapted species tend to be relatively big to facilitate efficient handling of cold—these reindeer are now the size of pronghorn. Their survival in brutal northern climes is accomplished by layering on gobs of fat. Since they have no predators and thus little incentive to run, they can afford the gains in adipose tissue.

Muskoxen also adopted the strategy of adding layers of fat. But because they still live with predators, an unlikely trade-off occurred. They now most often stand together with their helmeted armament dominant; running away rapidly is not easy with a body layered in fat. The accumulation of stores of fat might have been the cause, or the consequence, of this behavior. That part is not clear. What is clear, though, is that this is a good, though not infallible, defensive behavior.

Half a century ago, Danish field biologist Christian Vibe appreciated the consequence of the putative invariant design in herd security: "In Europe and Asia the Musk Ox became extinct, while the reindeer survived. The Musk Ox had no defence against Man and Dog, while the reindeer could escape by flight." Vibe noted that the ability to flee benefits some species but not others. Stand and defend were what muskoxen did most often. It failed them, though, when hunters arrived with their dogs. It failed them at times in the Paleolithic, and it failed them in the nineteenth and twentieth centuries when they fell to Danish, Norwegian, American, and British explorers.

Nevertheless, in remote areas without people, herd defense has apparently worked well, because muskoxen are still surviving, even with grizzly bears and wolves. A reasonable presumption might be that if armed and standing one's ground, death may be less likely than if fleeing. The fight or flight response, first described in the 1929 book *Bodily Changes in Pain, Hunger, Fear, and Rage* by Harvard physiologist Walter Cannon, brought acute stress responses to danger to the fore. For me, the fascination is in learning how the differences in flight versus attack translate into survival.

Some of the best information is not from muskoxen, but moose, and specifically those from Michigan and the Scandinavian Peninsula. Moose that do not flee enjoy a three- or fourfold difference in survival. So, if a standing defense works better than flight among moose, is the same true for muskoxen?

The answer: no one knows.

Interactions between wolves and muskoxen are infrequently witnessed, and trying to re-create the outcomes by snow tracking, as is done with moose, is rarely possible in the windblown and inaccessible Arctic. Some insights about muskoxen that do choose to flee come from Canada's Elsmere Island. That's where wolf biologist Dave Mech reported that 14 percent of the chases of fleeing muskoxen ended in predation. What we don't know is what happens when circular defenses do not break down in the face of a predatory attack.

Who are most likely to survive wolf attacks? Calves and yearlings, the clear analogues of a football quarterback surrounded by its offensive linemen, or is it the adults forming the front line? Since it's easier to find bodies than to witness interactions, there is more information from the dead.

Across two decades of work on Elsmere in July, Mech found seven wolf-killed calves, a yearling, and five adults. Because the small bodies of young are easily consumed and disappear more quickly, there were certainly even more that had died from wolf attacks. While nothing is known about whether these kill rates applied to those that stayed and fought or those that fled, what is obvious is that the young are more vulnerable.

A useful picture of muskoxen death and life emerges from Polar Bear Pass on Bathurst Island in the high Canadian Arctic. Biologist David Gray summarized an impressive eleven-year data set from the late 1960s and 1970s. Of more than sixty births, 11 percent of the young died due to wolves and 50 percent did not survive their first winter. Gray didn't know whether wolves were involved in this 50 percent, but over four of the years, all reproduction ceased. No calves were born. Food limitation was an enormously important factor. Data on adult female mortality also suggested that food was a stronger determinant of survival than wolves during that time; 75 percent of the deaths were a result of starvation, with only a third due to wolves.

For adult males, differences were in the opposite direction; 63 percent of deaths were due to wolves, perhaps because bulls are less likely to be in defensive groups and because they deplete more fat as a consequence of intense rut-related efforts to mate females. On at least four islands to the west of Bathurst, adult sex ratios were skewed, with more adult females living than males. This finding is consistent with the idea that post-rut males perish because of dwindling fat reserves.

Such interpretations are tricky due to an obvious bias. That is, it's easier to confirm the status of marrow fat, and hence condition, than to identify predation as the cause of death. The winter situation is even more complex because animals can grow lean and perish with no predation involved. However, malnourishment stress can also weaken individuals, leaving them more likely to be preyed on. Then, do we correctly infer that the muskox skeletal remains found with the bone marrow fat depleted died from malnourishment? Or might we wrongly conclude this when, in fact, the cause of death was predation on weakened individuals? In biological parlance, such deaths are compensatory since the enfeebled were likely to die anyway. Detecting cause of death is especially difficult when carcasses are old and hides crusted and where evidence of predatory attacks can be confused with scavenging or bones. Partly because of the challenge of detecting causes of mortality, unraveling the interaction between weather and predation is formidable; yet it is fundamental in order to understand broader impacts of climate.

Comprehending the relationship between climate and disease, as well as the parasites that carry many of these diseases, is also important. Where hosts become more susceptible to effects of warming and seek cooler climates to the north, they bring parasites or disease northward with them. This topic whets Susan Kutz's appetite. Energetic and creative, Susan doubles as a disease specialist on the veterinary faculty at the University Calgary. She's found that the caribou that migrate back and forth between tundra and forest edge are the likely conduits for the recent transport of lungworm to muskoxen on Canada's two largest western Arctic Islands, Banks and Victoria. While lungworm doesn't kill directly, it weakens immune systems, and both muskoxen and caribou then face additional pressures. Susan explains: "Many of the wildlife species in the Arctic are key for the Inuit and the Dene people. They depend on these animals for food and the practice of harvesting these animals maintains cultures and traditions that would be lost otherwise." Muskoxen are apparently more appreciated by indigenous populations in Canada than in Alaska. However valued in different parts of the Arctic, we still have much to learn about muskoxen survival. How does a changing climate affect disease, food availability, and predation?

Back on my little patch of permafrost, I still can't hear the helicopter but I see a bird in the distance that appears tiny. The wind is strong, muffling all sound. As the bird grows closer, rotors replace wings in my line of sight and a voice squawks on the radio, "The first dart missed. The second was good. We'll be right there to get you."

The ambient temperature is −14°F. I'm frozen. But then, in a matter of minutes, I'm aboard the chopper, and we return to the back side of the hill. The group is gone.

We set down to work on the immobilized female. I cover her eyes; they hold deep secrets. I imagine wisdom. I want to know her world—what she sees, thinks, and feels. I hold her head high. The horn boss is warm; heat faintly escapes. Her hair is thick and luxuriant.

A quick glance to the side is unsettling: her herd mates are a mile away and still moving. We'll worry later about reuniting them with their sedated member.

Layne draws blood and measures body fat with ultrasound. I fasten her collar snugly, then note the linear distance between eyes—seventeen centimeters. All the data are necessary for later assessments of my photo-imaging project. We check teeth; two incisors are permanent, the third deciduous. One of the canines is broken.

I insert my fingers into a plastic glove and prowl her heated bowel. Warm mucosa squeezes against my two cold digits, the heat providing a moment of sensory bliss. I probe another centimeter inward and emerge victorious with a dozen slimy nuggets.

We hoist the tranquilized cow onto a scale. She registers just over four hundred pounds. We load gear and are in the air even before she's up and mobile. We need to coax the fleeing herd back to the downed female. We're unsuccessful and can't change the herd's panicked direction. Then, Rick sees that our collared cow is now mobile.

"Shit. She's going in the wrong direction." We gape as she dips down a hill and can no longer see her group. This isn't good. With a sharp 180° turn, we veer and leave the area, hoping to disturb no more animals.

Fairbank's average winter temperature is a stunning 8.3°F warmer now, on average, than it was in 1949, with wild swings accompanying the changes fully underway in the north. In Siberia, lakes across the permafrost now crack under freeze and thaw cycles—something that never used to happen. In the past twenty-five years, 11 percent of more than ten thousand fresh water bodies larger than a hundred acres have dried up, the majority along the warming southern edge of the Russian Arctic.

It isn't only inland cities, like Fairbanks, or freshwater lakes feeling the heat. Coastal villages on barrier islands are feeling it as well. Alaska's Shishmaref is one and Kivalina another. With fewer than four hundred residents, the latter is the sole peopled spot along a lonely 125-mile coastal stretch between Cape Thompson and Kotzebue. Both villages have maritime climes with fierce wind and fog. "Kotz," "Shish," and Kivalina average eleven to thirteen inches of precipitation annually. Inland, the climate is generally colder, but because of winter inversions it can also be warmer and sunnier. Warmth is relative, and temperature can reach −50°F; wind chill can be as low as −100°.

With warming, vegetation changes, as do wildlife communities, all with consequences in humans' chase for foods. Migratory caribou, a long-standing source of nourishment for indigenous populations, are everywhere, although they had been absent for a while on the Seward Peninsula. In 1892, domestic reindeer were imported from Chukotka in Russia. The idea was to stabilize food security for Alaskan villagers; over four decades, herds grew to more than six hundred thousand. Sami herders from Norway were brought to Nome, where they taught herding to local Inupiat. Reindeer also escaped, which created feral herds that later hybridized with wild caribou. And over time, reindeer herding became more problematic and less profitable, so today, wild caribou are the mainstay.

Fred Goodhope Jr., the contemporary treasurer of the Alaskan Reindeer Association and longtime resident of Shishmaref, recently offered thoughts on food and climate: "It's the early melting and later freezing of rivers and the sea that are killing us. We can't get no caribou because the damn thin ice is too dangerous now. Getting across the ocean to Kotzebue for sheefish is tricky."

Scientists would agree, using the two-word expression "Arctic amplification." The concept is easy. Ice and snow generally cool Earth's surface by reflecting the sun's energy back up into the atmosphere. Warming temperatures soften, and then melt, snow and ice. Darker surfaces are exposed. Once this happens, the exposed earth absorbs more heat. With its ice coating gone, the sea also grows warmer. The effects are thereby amplified.

Wild swings in precipitation and warming temperatures affect more than food security for humans. Known as ground zero for climate change, Shishmaref is literally slipping away as higher storm surges pound the tiny island's three-square-mile surface. Kivalina feels the same erosive forces. As a first line of defense, sea walls have not worked. Neither have the $300 million requests to the U.S. government to relocate the two villages to higher ground. In 1964, Bob Dylan belted out the song "The Times, They Are a Changin'," referencing social issues. But the issues facing the people of the Arctic go well beyond social injustice to encompass matters of ecology, subsistence, and perseverance.

It's April 2010, and with the windchill, the temperature is about −20°F today. I've got thirty minutes until frostbite sets in. An ice fog descends. There is no visible helicopter in sight, and there is no radio contact.

We've now darted more than fifty muskoxen. I'm better prepared, this time around. In addition to the two-way radio, satellite phone, and GPS unit, I now carry two additional satellite phone batteries, an emergency flashing unit, an emergency locator beacon, and a ground-to-air radio. Still no sleeping bag or tent, though. If the helicopter goes down and if I can find it, there will be two bags aboard. There are three of us.

The fog lifts. Twenty muskoxen are panicked by the helicopter above. They thunder along a thin ridgeline. The problem is that I'm also on that narrow ridge, and standing in the only passable spot. The herd must quickly decide which is less dangerous, the lonely human or the ferocious machine. They select me.

A bull leads the charge. He grows larger, seemingly in an instant. I'm in their path—fifty feet away, no: thirty—and they're still coming.

Eight hundred pounds with upturned horns race my way, as the bull pilots their advance with no apparent thought, just momentum. His death wish is not mine. My mind spins, adrenaline pounds. Will Rick's prized door shield me from brute force? I don't think so. I drop the door. The swarm closely passes the door. And me too.

Layne's dart is perfect, and the chopper backs off. The herd mills around its fallen member. I retrieve Rick's door and find just a small nick. One more animal downed, weighed, measured, checked, photographed, and collared. Another day, another datum. Our total sample numbers fifty-one.

During the first year of restricted darting, I noted something interesting. The incisor teeth of three Krusenstern females were cracked or broken; three from the Bering Land Bridge were normal. Now, with another forty-five animals handled, the pattern remains. Ninety percent of the northern females have broken teeth, less than 10 percent from the south. The northern population has arrested demographic growth; the southern population continues to grow. Intriguing.

If the full battery of teeth increases chewing efficiency, then our first clues as to why the populations vary in trajectories may have emerged—dentition may be the cause rather than climate, unless somehow climate governs dentition. Females from the Krusenstern realm weigh less, a difference consistent with our expectation that missing teeth reduce feeding proficiency. Body fat is also less in the northern animals. One additional consistency is evident: fewer calves are counted relative to the number among northern adult females.

There's also a troubling inconsistency. If nutritional standing is mediated by masticatory efficiency, then why are the pregnancy rates in the two populations similar? I'm unsure. At this juncture, results are suggestive that somehow either food acquisition or food quality might explain the variation in body stature between the two populations.

Like sublimating snow levels, the differences in body mass between populations evaporated over the next three years. Females at both sites were equal in size and mass despite the consistency of patterns of dental intactness in the south and breakage in the north. Maybe perfect teeth are not necessary to survive, though it seems they should help. What also puzzled me was the variation in number of calves between regions.

During our darting expeditions, excitement came from more than muskoxen. From the air, we'd see Dall sheep, their white bodies a stark contrast to windblown slopes free of snow. Other times, there would be moose. Sometimes it would be caribou or flocks of ptarmigan in view, occasionally a snowy owl, a gyrfalcon, or a golden eagle. Sometime we'd see a herd of caribou on the sea ice. Twice we saw moose there and, once, a wolverine thirty miles offshore.

Seeing grizzly bears was a special experience. In early March, when temperatures could be as low as −17°F, solitary males were out; sometimes, females with cubs were visible. Even at winter emergence in March, bodies rippled in muscle, with fur turning creamy golden in early morning light. Simultaneously menacing and hypnotic, the magnificent bruins have an allure all their own. But my pleasure in seeing them soured as the helicopter chased the unlucky animals, sweeping in low, fast, and furiously, sending the petrified bears fleeing uphill or down. It was pitiful to watch them repeatedly dive head first into the frozen tundra, an act sadly reminiscent of an autistic child smashing his or her head against a wall. As Rick and Layne laughed, our aerial torment continued for longer than was kind.

Muskox herd mates were often loyal and quite reluctant to leave their immobile companion to the enemy—that is, the people in the helicopter. Once, a group of three males refused to back off an immobile female as we swooped twice to within fifteen feet. We then came in hard, creating a massive snowy wash. The herd finally departed, and we collared the tranquilized female. A different time we fired a .44 magnum above their heads to try to get them to scatter. They didn't respond. A second shot had an identical effect; that is, no effect at all. Finally, they ambled away. After collaring that female and getting her back on her feet, we returned to the air. In her post-drug stupor, she ran away from, rather than toward, the group. Once again, we hoped she would later find them.

That was the problem. We were never sure that animals returned to their groups. Over the course of the project, more than 215 animals were darted. Most returned to a group, and we felt confident that more than 90 percent found their original herds. Many reunited on the spot. But not all.

Projects like this can entail this sort of tragedy, in which a sensitivity to social dynamics of the animals studied is often sorely lacking. Because of Arctic logistics, it's perfectly understandable why research is aerial based and why behavior is not frequently studied. Yet individuals and groups share biologically nuanced relationships, and these affiliations should not be ignored.

Zebras and wild horses have strong bonds within their social units. So do lions, wolves, baboons, and gorillas. Despite their large herds, even bison form connections. The elephant story is well known. Why, or how, would muskoxen be any different?

Trauma is rife when social bonds are severed. Group life involves learning and socializing. It relies on communication and leads to the development of motor skills, sometimes through play and sometimes through fighting. It assures recognition of signals and of individuals and increases the ability to deal with stress. Group life might guarantee that a repository of information is handed down from one generation to the next.

For elephants, social learning is central to understanding danger, knowing where feeding sites are, and locating reliable water during drought. Among solitary species, bonds between mothers and young shape the acquisition of knowledge. Young moose discover where to migrate. Those who lose mothers do not. Yearlings learn from their moms when to run from humans and when not to, or the differences between bears and humans. This sort of knowledge transfer is vertical, from mother to offspring, whereas horizontal transmission is more frequent in groups.

Mechanisms of learning in muskoxen and other Arctic species await discovery. We do know that when herd mates snort, groups bunch; also, we have observed that youngsters can distinguish important members of their herds, which became evident during my photo-imaging visits to University of Alaska's Large Animal Research Station. We also know that the youngest among the herd are the most likely to respond to new cues—only they, for example, pay attention to ravens and occasionally chase them. While in captivity, muskoxen, irrespective of age, may ignore human voices, but some know when and by whom they're going to be fed, so vocal and visual recognition of humans can occur.

One of the more unusual interactions involves the responses of calves to disturbance. If human keepers stomp about or make loud utterances, calves run to their human mentors, undoubtedly perceiving the heightened response as one of danger—maybe even of a herd mate being attacked. The calves have no prior experience with danger; their reactions are something more deeply rooted.

Learning occurs in varied ways, among them through curiosity. In 1911, Welsh poet and sometime American hobo William Henry Davies, wrote, "A poor life this, if full of care, / We have no time to stand and stare." While Davies was commenting on impoverished humans, those lines make me think of juvenile muskoxen.

Sometimes I'd lie down on the tundra pretending to be a rock. It was almost always the youngsters who investigated. Two-year-olds were the most inquisitive, leaving the herd and approaching to within ten yards. Curiosity comes with benefits and with mortal costs.

It's not uncommon for some animals to stare at and then approach novel objects or dangerous carnivores. In the Serengeti, one researcher estimated costs of curiosity. During more than four hundred interactions between Thomson's gazelles and cheetahs, the young were killed ten times more often than adults when they stopped to stare. As we know from humans, teenagers take more risks, and these can turn costly until lessons are learned. Those that survive to adulthood may be the more prudent individuals. Natural selection is at work, as it always is, with the culling of the innocent or the careless. Being curious can be a good thing for learning unless you die as a consequence.

No one seems to know what muskoxen females see, experience, or truly gain by their membership in groups. We presume it's protection from predators, a reduced rate of burning fat, perhaps warmth, and maybe the beneficial access to cratered-out foraging sites for lichens or grasses.

If one desired an experiment to find out, it would be this. Detach a female from her band. Force her to solitary existence. Do it in winter. Add bears as they emerge from their winter dens. If she's pregnant, assess whether her calf survives. Replicate. That makes it real, if not cruel, science.

No one would seriously do this, not with a hope of passing the watchful muster of a state or federal animal care committee. And we didn't. Well, actually, we did: we did exactly that. We didn't purposefully separate animals, but our darting operation produced an unintended test in which we detached some females from groups.

It's not something I was proud of. While Layne and I had taken every reasonable step to assure psychological well-being and to return females to the security of the herd, we sometimes failed. I consistently questioned the trade-off between ardent science versus what animal ethicist Marc Bekoff calls compassionate conservation. It was a torment that would change how I'd conduct later research.

Unlike the predator-free caribou of Greenland and Svalbard, female muskoxen are hardly ever alone. Over eight consecutive years in Alaskan Beringia, I never saw one alone, other than our inadvertent separatists. In two decades in the Russian Arctic, Alexander Gruzdev and Taras Sipco never saw one on its own either, neither on the Taymyr Peninsula nor on Chukotka's Wrangel Island. Neither did John Tener during his many years in the Canadian Arctic. David Gray and another Canadian scientist, Anne Gunn, both reported such sightings as events so rare as to be noteworthy.

The solitary females we had unintentionally created offered an experiment, an extreme one. We could inventively determine how habitat selection substituted for group security.

We discovered the first isolate on a cobalt blue April morning—or rather, Marci Johnson actually did the discovering. Seasoned by ski-tracking wolverines in Montana, Marci was intent on finding our collared female muskoxen. While skimming the tundra in an airplane, she monitored her telemetry gear, listening for radio signals. With only a few hills and numerous lakes below, a large black object should stand out against snowy permafrost. And the radio signal blasted in Marci's headset. Although Marci was eagle-eyed, her pilot Eric Sieh's sight was even better. Still, neither could glimpse the female. The closer to the collar one is, the louder the signal. Given that the signal was now blaring loud as a Mick Jagger tune, both were perplexed. Where was the missing female?

Early legend had it that mammoths were a subterranean species, escaping Arctic cold by going underground. Indigenous peoples did this, at least partially, situating their abodes into embankments to block wind and retain heat. Up until at least the 1950s—by which time the first Russian and then American astronauts were preparing to travel into space—some Arctic inhabitants still eked out a living in semi-underground shelters. Mammoths certainly must have. Why else would their tusks and skeletons be discovered underground?

Could our elusive collared female be alive under the snow? A marmot probably would be. Muskoxen—no way. At least not according to 150 years of written observations.

The plane turned sharply. In the smooth snowy surface that stretched across an otherwise seamless landscape, a crevice dipped into the frozen earth—a snow hole, some eight feet across and seven deep. In the roofless fissure was a jet-black object against a steep-sided wall. Was our detached animal stuck? Had she fallen in? Was the collar malfunctioning, offering a false signal of an animal alive when she was dead?

Eric came in lower. The animal pressed her butt hard against the steely white platform she stood on. She was alive. Because of recent snow, there were no tracks. And because her collar was not GPS, there were no stored data. I didn't know how long she'd been in the snow hole. And we were helpless to do anything for her. I felt terrible.

Days later, we discovered a different solitary female. Again, a snafu—situation normal, all fucked up. Over and over, and throughout different years, we'd see things we just had not expected, and these included solitary females situating themselves in snow holes. Some individuals moved from snow hole to snow hole, ruling out the possibility that this was just an obscure chance event. How they located these bathroom-sized wind-whipped snowy depressions, so rare and unreported across the vastness of the Arctic landscape, we wouldn't know. Nor would we know the psychological effects of the social castration from the herd.

Six in total used snow holes in the Bering Land Bridge, an area with more lakes and rolling tundra, and with fewer crags or hills than the northern study area. Unfortunately, our unsettling actions repeated themselves between Cape Krusenstern and Cape Thompson.

Of the three loners in this more rugged northerly terrain, only one adopted the marmot's subterranean tactic. Instead of snowy fissures, the other two showcased their goat ancestry, isolating themselves on rocky pinnacles. One remained attached for weeks to a miserly spire a few hundred feet tall above a valley floor. The other, a two-year-old, was fixed to a more rugged spear that jutted sharply up, as if a staircase. Like the seven orphans years earlier, we couldn't determine this one's fate because she wore no collar, having become separated in our darting operations that targeted an older female.

I agonized about the impacts of their forced social isolation. Individuals experience anxiety, of course. Separations from comrades, as well as death, cause unusual behaviors among the living. Outward expressions telegraph feelings.

In humans, for instance, smiles generally connote one thing, tears another. In dogs, hackles and play bows reveal the difference between fright or aggression and an intention to frolic—differences noted as long ago as 1872 by Darwin in his book *The Expression of the Emotions in Man and Animals*. The behavior of wild animals can similarly reveal passion or despair.

Orphaned elephants may awaken with high-pitched cries after their mothers are slain. Elephant moms who have lost their babies stand for days with their heads down and ears drooping. Moose also uncannily express feelings. I once watched an orphaned calf run more than half a mile and then return to its mother's bones when chased; sadly, maternal protection would not come. A different time, I watched a moose return to the site where her calf was killed by a car. Whether social or asocial, some mammals remember the dead.

The fact that trauma causes neurophysiological change is well known in species like chimpanzees, gorillas, and humans. The mechanisms for those changes are not. Two areas of the brain, the cortex and hippocampus, are suspected to be the regions responsible, and these may be reduced in size after massive distress. We would never know the emotional toll of becoming a muskoxen orphan in the Beringian wilds, or that of the females we detached from their herds, though some clues were available from which we might learn.

Some secrets can be gleaned without more helicopter harassment, darting, and handling. I would need to return to snow holes and mountain retreats, find scats, and examine them for stress hormones. By extracting metabolic by-products from the frozen feces of solitary females and females in groups, it's possible to infer how sociality, or lack of it, affects their psyches. I'd feel better if I could ascertain the consequences of our disruptive captures, even if the knowledge did not immediately benefit the animals themselves.

A diagnosis of stress would also bolster my pursuit to understand how climate and other factors affect muskoxen, especially long-term behavioral stressors that could be physiologically costly. Cortisol, for instance, a steroid produced from the adrenal cortex, can inhibit individual growth, suppress the immune system, and modulate reproduction. If the subdued population growth from our northern study areas resulted from animals having experienced more stress than those in the Bering Land Bridge, it should be reflected in higher levels of cortisol in that population. But even if this were to prove to be the case, disentangling causes from effects would not be easy because stress could lead to failed reproduction or vice versa.

Scientists can be coldhearted or compassionate. And while doing science, one makes choices. Ethics and welfare matter. The idea of gaining information at any cost is not the part of research I wanted; I did not want to chase grizzly bears or muskoxen by helicopter. I would instead pursue noninvasive approaches more intensively. It was time to get on the ground, locate muskoxen again, and find feces.

A different set of logistics would then keep me busy—sleds and snow machines, crossing fissured and fractured ice, and finding biological technicians, or bio-techs, for short. I needed local guides. I'd begin by querying Tony Weyiouanna. He delights in telling everyone that he and his family have lived on Bering's land bridge for several thousand years.

> We saw two small boats being paddled from the shore to our ship . . . the wings of [one paddler's] nose on both sides were pierced by fine pieces of bone.
>
> GEORG WILHELM STELLER, *Journal of a Voyage with Bering, 1741–1742*, describing first contact between Europeans and native Alaskans, Gulf of Alaska, 1741

4 Where Worlds Collide

I'm in Shishmaref looking for a guide. "Jimmy knows snowgos," offers Tony Weyiouanna, mustachioed and graying. "He'll find the animals you want." I don't know Tony's brother Jimmy but I want to. I try to meet people before I work with them.

I check the general store. Owner Percy Nayokpuk has not seen him. "Maybe Curtis knows." But Curtis Nayokpuk is not around. Before departing, I take stock of the place. Shelves are half empty, with only a smattering of tins of Campbell's soup and peaches and boxes of Cheerios and Fruit Loops. The jam is next to containers of Folger's coffee and peanut butter. I know why caribou are needed for the table.

Bottled water is eight dollars a gallon. With snow and ice outside, why buy water? One answer to that quickly comes to mind: yellow snow. There is another option, but it doesn't provide the convenience of a 7-Eleven store, the nearest of which is more than four hundred miles away. To get good ice, an Arctic dweller needs a chain saw that works, fuel, and a functioning snow machine. He or she needs to ride ten miles to a lake, cut ice blocks, put them into cardboard containers, and ride back. At ten miles per gallon and with gas at $6–$7 a gallon, the option isn't so rosy. Ice then needs to be melted, but neither oil nor wood grows on the tundra. Fuel is expensive. The Arctic is a costly place to survive.

My search for Jimmy must continue. I open the store's door to exit. Mistake. The wind sends my hat sailing down the street. A woman in thick garments shoots by on her snowgo, followed by a twelve-year-old with no face mask and riding even faster. Fish, frozen in the cold, hang on a drying rack next to the hide of a grizzly bear. Seven skinned seals are, fittingly, laid out at the cemetery.

Someone tells me Jimmy is in the hospital in Nome, something about a bar fight. "This is a rough place," says Fred Eningowuk. "Eskimos get beat up."

A different guide, mentioned to me by some locals, catches my attention. He's a quiet man who spends time in a ramshackle cabin eighty miles up the deserted coast. He's shot polar bear and walrus. Last year he brought in thirty-five dressed out caribou for the elders of Shish. *Robin Hood? Or the gunslinger John Wesley Hardin?*

I investigate more. The man recommended is Fred Goodhope Jr., the treasurer of the Alaskan Reindeer Association and former reindeer herder I talked with a year earlier. An ex-bush pilot, he's now returned to his boyhood haunts beyond the volcanic remnants of Devil Mountain Lakes. It's a place he hunted with his dad. One night in the 1980s, his snow machine flipped about fifteen miles north of Shish. He had not seen the snow hole. Disoriented and with the temperature at about −10°F, Freddy walked through the night, at first warming with the movement. His clothing then froze. His toes and fingers lost feeling. In severe pain and hypothermic, Fred Goodhope Jr. hobbled into Shishmaref.

A few years later, Albert Weyiouanna—one of many in the Weyiouanna lineage—was hunting wolverines beyond Kotzebue. His snow machine broke down. Everyone's does. Or it breaks through ice, or becomes mired in deep snow, or turns over. It was early March, still really cold and too far from town to walk. The cable for his snowgo throttle had frayed. No fixing it. No spare. With no throttle, there is no gas flow, and no motion. This is when people die a miserable death. Or, it's when ingenuity kicks in.

Many from the "Outside" would have no ingenuity to offer. I wouldn't. But Albert's night bag had dental floss.

The Survivalfirst website (now defunct)—something that would not have been available to Albert for another quarter century yet—offers advice on how dental floss can save lives, such as being used to patch equipment, for fishing and sutures, or for making rope or snares. Albert—who had the powerful trait of common sense—didn't need the advice. Because he also had the benefit of well-seasoned cultural and traditional ecological knowledge. The first survivors of Beringia had their own version of dental floss: tendons and ligaments, which were used to fasten large arrowheads to the wooden shafts of spears made for hunting on land and sea. . He connected the throttle to the frayed cable with a length of floss, stitching it into a tight patchwork, and yanked. It was only a sixty-mile night ride to Kotz.

Alaska is rife with remarkable stories of survival and death. Jon Turk crossed from the Sea of Japan and paddled his kayak to the Alaskan coastline, first enduring the Russian navy, then orcas, brown bears, and walruses. Mike Horn navigated the entire Arctic Circle by foot, sled, and skiff in an amazing twelve thousand–mile feat, arriving in Nome after traversing Canada in winter. Senator Ted Stevens was killed when the small plane he was in crashed into a brush- and rock-covered mountainside. Christopher McCandless, the subject of Jon Krakauer's book *Into the Wild*, sheltered in a school bus but eventually starved. More than ninety-five climbers have died on North America's highest peak, Denali. Biologists surveying polar bears, caribou, and Dall sheep perish in accidents or just vanish into the depths of the Beaufort Sea. Extreme adventurers, hikers, and 'bilers disappear. In 2010, a high school teacher was slain by wolves on her morning jog outside a remote Aleutian village. Grizzly bears, avalanches, drowning, and hypothermia have killed a lot more. Few people have a death wish, but Alaska has a special draw for those toying with potentially fatal adventures.

I wanted to learn from the likes of an Albert or a Fred. I needed someone who had experienced the rhythms of the land and who hunted; someone who enjoyed tracking a hare across ice in a snowstorm; and someone who didn't need a GPS to know north from south when the wind is gusting at fifty or when visibility stops at the edge of one's nose. I wanted a mechanic with dental floss savvy, one who could replace the belt of a snowgo, pull its engine at −10°F, or fix a decrepit Coleman stove without parts ordered from a dot-com like Amazon. If they'd work with me, I'd watch and listen. I'd learn. I needed to function in a place where I had no primal roots.

Species come and go. Animals vanish from one place but still survive on continents elsewhere. There is the world's most northern antelope, the saiga, which died out in North America but persists in steppes and deserts of Mongolia and Kazakhstan. Or take North Africa. Like Beringia, it appears to be at or near the juncture of continents. African lions made it to Europe and Asia and, finally, North America long before humans. Hyenas also colonized Europe and Asia. Horses spread in the other direction, from North America to Africa. Ask anyone to name survivors of North American horses in Africa, and they'll get some right—zebras. The less obvious group would be their stripe-less relatives, the asses. Then, it gets more complicated. Ask again, but this time about Africa's modern bears and deer. There are no bears or deer there, right?

False. Until the nineteenth century, the Atlas Mountains of Morocco and adjacent Libya had brown bears. Red deer, also known as the Barbary stag, is Africa's only native cervid, and once spread from

Morocco to Tunisia. Until recently, Africa was much like a small bit of Europe and included red deer, wild boar, and brown bear. The losses of these European denizens across a swath of former ranges in Africa may have generated ecological openings. We just don't know.

Changes that more recently preceded us are usually far from noticeable. No living person in southern California has seen pronghorn in the basins now dominated by freeways and millions of people. It's unlikely any Californian alive today saw grizzly bears in California, as the last was shot in 1922. What we see now creates a falsity if we naively believe it is even close to conditions preceding human habitation, even if the land were to lack a footprint as large as Los Angeles. Prior mammal distributions are both illustrative and deceiving.

North America once had true monkeys. Outside of Mexico, they're long extinct. The next primate to enter the continent was us—*Homo sapiens*. These humans from Asia were the original North Americans, the ones who established hunting as the primary form of subsistence. Entry onto this people-less continent was of course from the west, using the northern thousand-mile-wide land bridge. Others hopscotched small craft down coastal routes. The ancestral colonists found a land and sea fauna that was naive about hunting humans.

In truth, not all animals were wholly naive. The New World mammals, which had deep ancestries in Europe and Asia, had been hunted by Paleolithic peoples for thousands of years. Cave paintings depict armed humans, and rock cairns reveal sites that funneled ibex into a dead end where they were then bludgeoned. Decades ago, Scandinavian paleontologist Björn Kurtén suggested that "most of the European invaders of North America, the moose, wapiti, caribou, muskox, grizzly bears, and so on were able to maintain themselves, perhaps because of their long previous conditioning to man." Species that evolved exclusively in North or South America—the camels and their close relatives, guanacos and vicuñas, or pronghorn—might not have the ancestral savvy to assure persistence if Kurtén was correct. He wasn't. These species are still with us. So are the likes of bison, moose, and others that crossed the Bering Land Bridge.

The question is not whether humans killed animals, as obviously our ancestors did with exacting efficiency, and as we still do. It's whether we drove the extinction of large prey at such a rapid pace that they could not adapt. This idea is called the overkill or blitzkrieg hypothesis; its chief architect is Paul Martin.

Martin felt that the most recent extinctions were not due to effects of climate but to humans. He noted that within a mere speck of time, the big game fauna simply collapsed after humans arrived. Nearly 75 percent of species larger than a hundred pounds vanished between nine and thirteen thousand years ago. They had been browsers and grazers—a half dozen species of elephants, llamas, and camels, two types of peccary, and three kinds of horses. Disappearing also were dire wolves, short-faced bears, and *atrox* lions—these three being nearly indistinguishable from contemporary species in Africa and Asia. Even the American cheetah, an animal most closely aligned with modern cougars, disappeared from North America, it's only known continent.

When species enter new lands, ecological changes inevitably occur. Modifications in Beringia, and subsequently North and South America, typified such community-level change as people crossed over from Asia. With the arrival of armed humans, a suite of species that once had only serious four-footed predators now had one with two feet. Such intersections create experiments and opportunities.

Much later, throngs of Europeans entered the other side of the New World. They first arrived in Greenland and then reached Canada's subarctic northeastern shores. The subsequent arrival of Spanish seafarers in the Caribbean in 1492 also led to the redesign of every ecosystem. Domestic animals and people introduced new pathogens. The inevitable intersection of cultures, between Europe and the Inuit at the northern edge of the world to the "Fuegians" (Yahgan) at the tip of the Southern Hemisphere, initiated the modernity that challenges us all.

If we wish to appreciate how a changing climate might affect species, we also need to understand how humans have had an influence on species, both directly and indirectly. Ecological baselines can help.

But baselines are complex. When is a good starting point and when is a good end? Which geographies are considered? Take, for example, the North American bison. It is best known as a species whose "natural" distribution was on the central prairies where its densities were highest. What fewer people know is that bison also inhabited Atlantic forests from New York and Pennsylvania to Georgia during the seventeenth and eighteenth centuries. This expansion eastward from the prairies may have been made possible by the decimation of millions of Native Americans from pathogen-borne diseases brought by the Europeans, which essentially emptied those forests of human predators. While ecological baselines are a human construct, which shift depending on what we view as starting points, they help us understand landscape changes. They're also valuable for disentangling influences of humans and climate on ecological systems.

The earth's climate, plus an element of chance, plays an enormous role in who lives, dies, and thrives. And where continents collide or connect, the great forces of nature are also at work. Tectonic activity sent the Indian plate on a thirty-five hundred–mile journey in which it slammed into Asia forty or fifty million years ago. The impact created the Himalayas. The uplift was fifteen hundred miles across and a hundred to a hundred fifty miles wide, and the only spot on Earth where peaks reach over eight thousand meters (or twenty-six thousand feet). The Himalayas intercept monsoons; generate their own ice, troposphere-like winds, and a massive rain shadow; and sculpt evolutionary and ecological processes. An even larger natural force are the oceans, which mediate the hydrosphere. The conflation of climate, water, land, and geo forces regulates associations between continents. These include the area between Asia and Alaska, the isthmus between the two Americas, and one that never connected Micronesia to Australia, an area just south of the world's deepest trench, the Mariana.

It is places like these where species reach extremes. Some fifty thousand feet lower than the world's tallest peaks—a full twenty-five thousand feet below the sea's surface—is an unusual fish. This six-inch-long species of snailfish, discovered in 2014 in the Mariana Trench, endures a greater hydrostatic pressure than any other fish in the world. A different sort of record is set by Rüppell's vulture. In 1973, one smashed into a jetliner at thirty-seven thousand feet—a height some eight thousand feet above the frigid summit of Mount Everest. In this case, the collision point was in Africa, above Abidjan, Côte d'Ivoire. And in 1926, a Lutheran pastor climbed Africa's highest mountain and discovered a dead leopard at about 18,500 feet, the frozen carcass of which Ernest Hemingway mentions in his short story *The Snows of Kilimanjaro,* written some ten years later. In Asia, pika, blue sheep, and wild yaks own altitudinal records that top out near nineteen thousand feet. Tibetan herders occupy the same realm. They possess nearly twice the concentration of nitric oxide in their blood as other humans, an adaptive trait for high elevation.

The word "extreme" has different connotations dependent on the context. For scientists it's a statistical construct, typically the high or low end of a variable. Meteorologists, for instance, use it to classify events like heat, drought, or the different sorts of storms that fall within some bounded normal range; those conditions beyond, such as the ninetieth or tenth percentile, are progressively more extreme. Dictionary definitions align well—intense or excessive, acute or outermost.

Other physical scientists talk of extreme environments, including those beyond the realm of humans; some other organisms can be well adapted to these. These may be areas of unusually high pressure, like five miles below the ocean's surface with its snailfish, or areas so dry, hot, or high that the only extremophiles are bacteria.

Among these are the extraordinarily odd "water bears," better known as tardigrades. They've been compared to pandas or pygmy rhinos, though with eight legs. These microscopic relatives of roundworms survive temperatures more than 300 degrees below zero and temperatures above boiling. Beyond bears—water, polar, and grizzly—there is yet another type, the Arctic wooly bear. Not ursine at all, this Arctic moth synthesizes glycerol to protect its cells from freezing and survives when temperatures drop to −60°F. Because its period of feeding is so truncated and dormancy so prolonged, completion of the life cycle from eggs to breeding can require seven years.

My interests are more pedestrian and narrow, and not just a focus on mammals, but those at the edge of existence. Edges can be more than extreme elevations or frozen limits. In formerly warmer climes, for example, giraffes and African rhinos reached their bounds where resources petered out. They, along with hippos, made it to the steamy fringe of what is now the Sahara Desert, where—prior to climate-induced drying—they thrived only a few thousand years ago. Far south in the Kalahari and Namib Deserts, all three persisted until just recently, when hippos along the Kunene River separating Angola from Namibia were hunted-out. In East Africa, elephants once lived as high as made it to thirteen thousand feet. In Bhutan, known as the land of gross national happiness, tigers go higher. Beyond Bhutan, in Russia's Amur River Basin, tigers still experience deep snows and bone-chilling cold; moose and red deer (elk) are their prey. Snow leopards, symbolic of high crags and remote Himalaya vistas, can range in the other direction, dropping low into dry juniper forests in Kyrgyzstan and Afghanistan. Or, in Mongolia where they've been found up high on the domed-shaped Sutay Glacier, they also drop below thirty-five hundred feet, elevations at which bison once roamed the prairies of Kansas. Some species are truly catholic.

My metaphorical snow oxen are less so. They're residents of cold, high, or often remote realms. Among key groups of snow oxen are the Caprinae and Bovini (both in the ruminant family Bovidae), whose members occupy on four continents that comprise about seventy countries. The Caprinae are pure climbers and withstand brutal extremes occurring in torrid deserts of North Africa, Turkey, and Mexico. Some occupy biomes in Vietnam, Taiwan, and Malaysia, or moister and more temperate climates, from the Swiss Alps to the Rocky Mountains; and others manage to reach higher and colder sweeps of the Hindu Kush and Karakoram. They've spread from the Tibetan Plateau to Siberia, and even onto the Kamchatka and Chukotka Peninsulas, to then cross to Alaska's Brooks Range and east into the Yukon.

In Latin, the word *capra* translates to "she-goat" or nanny-goat. The lives of most caprids are anchored to precipices, with the sole anomaly being muskoxen. As climbers of Earth's more recent geological uplifts (areas of mountain building during the last few million years or places of recent deglaciation), they exploit low-growing vegetation. Populations are often island-like, their mountainous outposts typically surrounded by seas of unsuitable terrestrial habitats. Populations are discontinuously spread; some pockets are of high densities, others low.

Within the Caprinae are three surviving tribes. True sheep and goats are called Caprini. These include argalis, the world's largest wild sheep. The bighorns and thinhorns (Dall and Stone's sheep) of North America, snow sheep of northeastern Asia, and ibex—ranging from Ethiopia and Yemen to Afghanistan and Russia—are all tribal members. So are blue sheep of the Himalayas. The Rupicaprini include mountains goats with their thick pantaloons and white-yellowish coats. These are truly mountain antelopes, at least in origin. So are the darker chamois and the small gorals and larger serows, dwellers of rugged mountains. The muskoxen and takin are often classified within a tribe called Ovibovini. This means part sheeplike and part oxen-like, and hence the fusion and confusion of Latin labeling.

Two other relatives, chirus and saigas, use open areas and escape predators by fleetness. Chirus also carry the moniker of Tibetan antelope, though genetically they're more goat than antelope. Saigas, once classified in a fourth tribe of the Caprinae, the Saigini—along with the chirus—are now known to be more gazelle-like, rather than being of goat ancestry. From a conservation perspective, both chirus and saigas were formally categorized as endangered due to recent population collapses.

Lastly, there is yet another tribe—the Bovini in the subfamily of Bovinae—whose specific family, like the Caprinae, is the Bovidae. Among the Bovini are the true Asian and African antelopes and powerfully built, cattle-like wild ungulates. It is these, the enormously muscled grassland grazers like African buffalo and North American bison and the lesser-known but similarly sized gaur and banteng of the Indian subcontinent, that many associate with an oxen ancestry.

Moving from steamy Asian lowlands of the gaur and banteng, up and over the Himalayas, and north to cold plateaus, one encounters an additional impressive bullock of the Bovini, the yak. So closely related to bison are yaks that hybrids were purposefully created in the 1920s when Canadian interests in cold-weather breeds trumped concerns about genetic dilution. More to the point, wild yaks are clearly oxen, but not an ox of the like confused with the ill-named muskox. They are more of the Blue Babe variety, one related to aurochs and native cattle.

Much early confusion existed about the identities of yaks, bison, and muskoxen. Late in the 1700s—about the time Britain's thirteen problematic colonies along the Atlantic coast were talking about sovereignty and Paul Revere was about to make his famous ride—a young German named Peter Simon Pallas emerged in Europe. He shifted his occupation from medicine and accepted employment with the Saint Petersburg Academy of Sciences. Between 1768 and 1774, Pallas explored steppes, mountains, and deserts of a loosely knit Russian domain that included the Altai and Transbaikal. Charles Darwin knew of his writings and quoted from an 1841 report of explorers crossing Tibet and deserts north: "According to Pallas [in 1777] the Mongolians endeavour to breed the yaks or horse-tailed buffaloes with white tails." This is an indication of the confusion concerning common names. In 1780, Danish missionary and zoologist Otto Fabricius wrote *Fauna Groenlandica*, a 468-page tome containing an observation of the horns, hooves, and hair of a large animal floating in ice near Greenland. Initially believed to be a yak that had drifted in from Siberia, Fabricius later corrected its identity to that of a muskox.

Today, we know muskoxen and wild yaks both possess unique but parallel features to survive Earth's most northern and highest peripheries. Each benefits from group living and has a long skirt draping to its feet; both sexes of muskoxen and yaks possess piercing armament. Though different in ancestry, and with wild yaks being twice the size of muskoxen, both are the snow oxen in my mind—extreme beings in harsh lands of cold and scant food, and where precipitation might be half that of the warmer Sonoran or Mohave Deserts.

I have a problem. If I want to understand muskoxen, I need to be in the field. I'll also learn more if I'm with someone who knows the Arctic better than me. Fred Goodhope Jr., the ex-reindeer herder from Shishmaref, does. He agrees to guide.

I'll now need three snow machines—for him, a field tech, and me. We'll need sleds to haul us hundreds of miles, ones that do not snap in brutal cold and whose shackles can withstand beatings against rock and judderings on tundra. The gear will include more than seventy gallons of gas. That's a big deal, because pulling five hundred pounds of fuel will burden pathetically inefficient gas-guzzling snow machines. We'll carry trunks filled with tents and sleeping bags, clothing and food, and pots and pans. There will be coffee and tea, frozen fruit, power bars, cashews and almonds, and chocolate. Cereals will be mixed with boiling water. We'll strap on thermoses. A winch, an ice ax, and shovel will be necessary. Blazo, the local name for white gas, will fuel our Coleman stove. Because the Arctic has little wood, we'll need to bring it and fill a primitive stove with the wood for warmth. Caribou skins will cushion the gear in sleds and offer sleeping comfort.

I finally sort out how to get the sleds' nine containers aboard a bush plane, and the other gear and I fly down to Shish from Kotz. Part of our payload includes containers filled with petrol, something that leaves the flight officials in Kotz without smiles.

Other issues require attention. Batteries die quickly in cold. I need many different sorts—for two-way radios, satellite phones, GPS units, three cameras, and torches. To keep them relatively warm, they're worn close to my body, but that comes with a cost. I'm uncomfortably weighted down. The National Park Service suggests even more gear—a wall tent (i.e., a tent with perpendicular sides) and generator, skis or snowshoes, lanterns, and even cots. We need to be self-sufficient, so this all makes good sense; people with inadequate supplies die in agonizing ways.

I ponder trade-offs between bulk and mobility and remember lessons learned from earlier adventurers and explorers. Mike Horn powered a sled with his two legs and skis while crossing Canada and Russia in winter. He chased adventure, not science, but knew about survival. In the1830s, Arctic explorer John Ross learned from the Inuit: "They could travel easier than we, could find delights where we experienced only suffering . . . could regale in abundant food. . . . We were out of our element, as much as in the philosophy of life as in the geography of it." Knud Rasmussen and Roald Amundsen each traveled light, adopting native ways. Sir John Franklin's search for Canada's Northwest Passage was the contrast: hefty loads. He and his entire team were laid to rest there.

Luck: Is that to be my ultimate hope for safety?

Perhaps Fred is my salvation, because he carries a legacy for luck. He rarely uses back-up gear and does not have a satellite phone. He once crashed his plane into the Chukchi Sea but was pulled out, wet, cold, and shaken. He seems nice enough. One goes with a hunch. I will risk it.

I also need someone from the north who knows the mountains, river valleys, and tabletops of our northern study area. I plan to fly to Noatak and later to Kivalina, both native villages of a few hundred.

I'll offer talks on climate change and wildlife. It's March, and the world is still bound in ice. Belugas are not yet migrating, seals are not abundant, and caribou are to the south. Most hunters should be around. If they attend, I'll have options and will ask about guides.

The villages have no hotels. I sleep on cots at the high schools at ten bucks a night and leave before students arrive. The venue for my talk will be the drafty community center.

Like in Kotz, coffee is set up and packaged cookies spread out, while I greet village elders in the hope of drumming up support. Thirty minutes before my presentation, I don parka and snow pants and walk narrow streets with snow piled six feet high. Fliers are stapled to walls at the post office and general store, places where worlds and cultures intersect. Beyond rural Alaska, events are advertised well in advance, often by social media. I was not "Outside" of Alaska; instead I was outside in the cold.

"Why are you here?" asks a man, smoke swirling from his cigarette. Despite my heavy dress, he knows I'm no local. Anthropologists working five decades earlier in other coastal villages had regularly encountered similar questions. None were personal—just people unfamiliar with outsiders and wondering about their business. I felt the same. I didn't use words like "study" but explained I was about to talk about muskoxen and invited him in for cookies.

In Noatak, eight attended my talk; in Kivalina, three. One I interviewed about being a prospective guide. He asked me about religion. Missionaries had heavy impacts throughout Alaska. My interviewee, with eyes ablaze and a voice of scorn, said if he ever heard anyone dissing the Supreme Being he'd kill them. My search for a guide continued.

It's an April morning when the plane departs Kotzebue and drops us in Shishmaref. We're finally ready to go find muskoxen. With me is Gretchen Roffler, a seasoned Alaskan biologist. We all meet at Freddy Goodhope's place, which abuts a frozen lagoon.

"Maybe we can leave in an hour or two. It's still below zero you know." Fred's idea—to wait until it warms—makes sense. It wasn't even noon. Besides, we'd need to clock only twenty miles to a cabin on one of the inlets before dark.

The entry to Fred's home is strewn with shards of frozen fish and skinned caribou. A chain saw is out front, along with a rusted harpoon and whalebone. Reindeer and caribou antlers mix with hides. A muskox head is at his neighbors' house. Spare parts from dead snow machines and chains peek through drifted snow. His dog, part husky and part barker, is staked where cardboard breaks the wind.

It's now 4 P.M., and we're still searching for fuel and more jerry cans. The store is closed, and Fred can't find the key that Percy gave him for the gas pump. He also can't locate his tool bag. His heavy snow pants have disappeared. His rifle is ready but the extra ammunition is missing. I'm OK with all this.

It's called Alaska time. And it won't be dark for another four hours anyway. The temperature has risen to 6°F. The wind has gone up too. Even with our sheltering masks, the left side of our faces will feel more like −15°F once in the wind. I ask Fred why he won't wear one. "I'm adapted," he laughs, and asks, "Are you both finally ready?" But, just as we're leaving, Fred stops his snowgo and runs to his shed. He reappears with a few chunks of caribou.

We reach the small homemade cabin with about an hour of light. The Coleman stove doesn't work. Fred can't fix it, so heads back to Shish. I ask he if wants to take a sleeping bag in case of a breakdown. "It's not that far." If the fog lifts, he'll just see us in the morning.

It dawns cold, about −20°F. The fog cuts visibility to a hundred yards. Gretchen and I wait, hoping to find the last GPS spotting of our solitary muskoxen cow, who hides in a snow hole. By 2 in the afternoon, the fog dissipates. No sign of Fred.

We struggle to start the snowmobiles, choking them, flooding them, and starting over. Although my mustache is frozen nearly solid, I'm now warm, as is Gretchen. The machines finally fire up and a veil of blue smoke overtakes us.

Gretchen grabs the pump action 12-gauge shotgun and slugs. I toss two sleeping bags and an emergency tent into a trunk and strap it to a sled. We zoom off to find Missing, the name of our solitary female muskox. We pass hoarfrost still heavy on willow and scare up ptarmigan. A wary moose mom and her yearling watch, reluctant to leave the river bottom. Arctic hare and red fox tracks abound. My GPS points to the southeast, and there are two and a half miles to go.

The sun warms the land, or so it seems. The temperature is 4°F, but it's calm. I scan ahead, looking for signs of Missing. Heat shimmers off the snow.

I don't know the protocol, or how best to approach. There is no literature on this. People just don't know about muskoxen in snow holes. Will Missing abandon her refuge that is so like a marmot's? Will she push her rump against the sides of the cavernous wall, using it as she might use herd mates, daring us to come closer? Might she charge?

Maybe Missing had left the hole. If she did, we'd see tracks, right? I ask Gretchen this, trying to assure myself that I am rational. What if she is in the hole? Is the next step merely a stealthy approach, to then peer down from seven-foot vertical walls and hope they don't cave? If they do, my death would be swift, inside a muskoxen trench awaiting a fate far more serious than the blows of Mike Tyson when pressed into a corner.

We kill our snow machines a thousand yards shy of the alleged snow hole. Visibility has improved. The sun hammers down, and it's 8°F. There's barely a ripple in the white landscape, certainly not a gaping hole. No muskoxen. No tracks. Doubting the credibility of the location, we restart the now warmed machines, and go closer.

Within two hundred meters, a few ridges appear, but no tracks. We stop the machines and walk. *There just can't be an animal here.* Just to humor myself, I ask Gretchen to grab the shotgun. I'm doubtful. *There are no tracks. There can't be a snow hole.* I lead.

Gretchen trails, nervously. I move in alone. At twenty yards I detect a crevasse; no—it's a cleft, sinking into the frozen earth. My heart thumps. I can't think. I close the distance. My eyes focus on a single spot. I don't know if Missing is in there.

I will myself forward, a silent step followed by deep breaths. The sun beats. I drop to hands and knees, crawling. In front of me is a depression, a deep one. I'm at the edge.

Tracks tell secrets, not lies. If one ever wondered about the social lives of long extinct species, seven million–year-old footprints in the Arabian Desert reveal that relatives of contemporary elephants lived in herds. A dozen of the portly beasts walked together in what are now bone-dry but preserved riverbeds.

Sand and snow also disclose truths. Damara and San hunters of southern Africa track ostrich across desert gravel. Alaskan hunters do nearly the same. Riders on snow machines follow the clawed prints of grizzly bears on gritty tundra until the padded trails turn to bruins, and the terrified animals are shot. The dots of a horned beetle's tiny feet or wriggling of a viper remain when sand is not shifting in wind. The delicate whirl of a single blade of grass endures in a light breeze. In snow, the mark left by fluttering grass is nearly identical to what might be left in sand. Pebbles leave trails when rolling down sand dunes. It's the same with snow. With wind, though, all such evidence disappears.

My friend Archie Gawuseb, a skillful Damara tracker, could do more than track rhinos in the Namib Desert. He could identify male from female by the imprints that remain from their gait. At the opposite end of the world, Canadian Inuit are said to also distinguish male and female polar bears by their gait. Tracks tell honest stories.

How can a muskox be in the snow hole? There are no tracks, and there is no wind to have obscured them.

I inch closer, and peek down the white wall. The snow shelf holds. This fifteen-foot long concavity has scat everywhere. Something is here. It's an arm's length away. *Holy shit.*

Horns curve upward, close. The sharp tips of them have my full attention. My breathing is deep. If the ledge gives, I'm in the snowy cavern. At that point, the muskox Missing becomes Mighty Mike and I become dead meat.

I must move back but risk destabilizing the shelf in doing so. I hadn't yet taken the shotgun off my shoulder—my miscue. I do nothing. But, how dumb is that? Can't Missing simply back out of her snow-encased marmot shell and attack me anyway? With her intoxicatingly close, I'm neutralized and can't pull myself free. We stare. I look down, she looks up.

Gradually the vapor lock subsides. I back away. She remains. An hour later, Gretchen and I rev up our snow machines. Missing backs out and leaves. We gather feces. I feel terrible. Not only had she been isolated from her social group, but we've now caused her to abandon her snowy safe zone.

Inadvertent though these occurrences were, I was deeply troubled by all the separations we had caused. Two of the unintentionally quarantined females gave birth in their snow chambers; one calf was alive, the other dead. The longer-term consequence of social isolation remains a mystery.

I had wanted to know what muskoxen see and feel and, essentially, to understand their perceptual world. Now, at least, with their frozen poops stored in baggies, analyses of glucocorticoid metabolites—stress, or cortisol, levels—will reveal whether females relegated to snow holes feel the loss of their comrades. Maybe they feel nothing.

It's late in the day when we reach the unheated cabin. The sky slips from sapphire to orange. Later, in darkness, aurora borealis will seize control.

A distant whir grows louder. Fred. His machine had problems. He had pulled the engine and replaced parts I'd never heard of. He bought new shackles for our sleds. He's frozen. We fire up the Coleman, heat the cabin, and discuss our different adventures as we prepare dinner. He didn't know about snow holes. Fred offers a caribou and bologna sandwich on Wonder bread. I boil curried noodles with dried chicken and veggies. We share. Tomorrow we'll target animals for photo data.

Over the next week, we chase tracks and GPS points. We haul heavy loads across wind-whipped ice and over permafrost. Fred leads as we drop into soft fluffy riverbeds below steep slopes, nudging machines down and over fragile snow bridges. We accelerate through overflow as water surfaces above ice. Despite face masks, the side wind is brutal. Sometimes I wear two. We cross the world's largest maars, landscapes where groundwater under permafrost was blasted upward by volcanic explosions and created massive, water-filled depressions; these refroze on the surface. Generally only a kilometer or less across, the ones at White Fish and Devil Mountain Lakes are herculean in size.

Some days there is sun, other days are pure shades of gray. Fred navigates without compass or GPS. Ground blizzards obscure all landmarks. With silent confidence he finds shelter, at times fifty miles or more away from where we stayed the night before. "It's hard to see out there today. Do you think we should stay put?" He laughs. We push off.

One target is Inuigniq (Cape Espenberg)—meaning the area of "no more people"—up along the Espenberg River at the northern tip of the Bering Land Bridge. The little abode that Fred built there sits just north of the Arctic Circle. Remnants of his defunct reindeer fence and a few scattered hides remain alongside, as do rotting barrels, rusted metal, and the skulls of polar bear and walrus. Whale ribs poke through snow along with caribou and moose antlers. Facing the Chukchi Sea is an old couch and an outhouse filled with snow.

The entry to Fred's cabin is also snow packed. Today, we're lucky. The drifts are only four feet. Inside, the ice-filled mausoleum glistens with frost; its thermometer registers −3°F.

We haul wood from our sleds and the driftwood we've scavenged along the coast. The cabin warms. Its iced inner coat turns to a steamy froth precipitating onto our gear.

It's still dark when I awake in the cold, cluttered cabin. Another year has passed. Fred is sleeping deeply in his bag on a caribou hide. My field tech is Wibke Peters, a German doctoral student at the University of Montana; her Bavarian efficiency is appreciated. Amid piles of gear, I move softly to light the stove. My coffee infusion is important. It's ninety minutes until we're ready to ride.

Fifteen miles pass until a group of thirty muskoxen appears. We craft a plan for our photogrammetry work. Wibke and I will approach on our two machines. Fred will circle from behind to steady the group. Once we've completed photo assessments, we'll measure snow depth and hardness. This indirect measure will be the frequency of steps per hundred in which an animal breaks the surface. Muskoxen are supported when the snow is hard. While snowpack has numerous layers that reflect melting, wind, and deposition, and different sorts of penetrometers offer sophisticated measures, we travel light. With constraints on time and groups to find, I accept the trade-off to gauge whether snow texture mediates an animal's decision to flee or remain.

We move in on foot as Fred's snowgo idles gently, a hundred fifty yards behind the tightly clustered herd. Wibke carries the laser range finder. At precisely forty-two meters we stop. My measures for photogrammetry are in meters. I remove my gloves. The four hundred–millimeter lens is bitterly cold. I peer into the viewfinder. Immediately, it freezes as the warmth from my forehead collides with the icy screen. I can't see the juveniles I'm to photograph. I scrape ice crystals with my nails. My finger temperature plummets, perhaps down to 60°F, an unpleasant feeling. The camera's autofocus doesn't work. Blowing snow doesn't help the laser. I'll hand focus even though my fingers are not nimble.

With pencil and data book in her bare hands, Wibke whispers too politely, "Joel, go faster, my fingers are cold," as pages flap in the zephyr. I still can't see through the frozen lens.

The herd starts to move off. In a perfect distraction Fred revs his machine. We pull on gloves and move a dozen steps. The animals look at us, almost with a willingness to cooperate. I hand focus again and call out data.

"Two-year-old, female, frontal, number twenty-three, twenty-four, twenty-five, lateral."

"Which two-year-old, the one on the left, or the one on the far right?"

"Far right, crap—where?"

"Oh, yes. Yes, left, I see it. I didn't get the distance."

"OK, OK."

The one on the left disappears back into the group. Adorable faces peek forth among crowding linemen. The herd is standing, not whirling. A large bull and three cows cluster tightly along the herd's leading edge.

"Yearling, front center, frontal—do you have it?"

"Thirty-nine meters. No, thirty-eight, shoot. Did you get it?"

"Yes," I happily offer. "Number 27 and 28—you got that, right?"

"No—you missed a number. Joel—isn't that yearling cute, hurry—get it. I can't feel my toes anymore."

"Stop your whining. We've got serious issues. The damn screen is icing again."

Our fingers lack sensation. Soon they'll burn painfully as blood surges. We finally return to our silent sleds. Fred knows we've secured our data photos. In a single 180° turn, he leaves the bunch. The animals just stare. We converge away from the herd and gather their feces. So goes the chaos, the harried missives between one another, and photogrammetry. This time it was twelve minutes to photograph each juvenile for frontal and lateral views.

It'll be another twenty vigorous miles until our next group. Bouncing over frost mounds and tortured tussocks, our bodies warm. Handgrips are heated. It's still plenty cold, but knowing that we can get on the snowgos and generate heat is a good thing. Each field season, I'll have covered six hundred to nine hundred miles.

Fred's Arctic backyard is changing—physically, culturally, and biologically. His dad had never seen a moose and didn't even know what they were. As the Seward Peninsula has warmed, riparian zones grow shrubbier, and habitat grows suitable for an array of different species. On first sighting a giant antlered deer, Fred Sr. thought somehow caribou had mutated. I enjoy absorbing indigenous knowledge and lifestyles and learn from nightly musings with all the guides over the years.

Fred describes trade school in Los Angeles, where he drank with bums on Alvarado Street. I grew up on Wilshire, only a few miles away. He talks of Los Angeles's diverse cultures and his Alaskan ancestry. I talk about days in high school, the many languages I was exposed to, and how the challenges of war, integration, and civil disobedience were reshaping America. Conversations shift to animals and money and lifestyles.

Modest as always, Fred talks of piloting bush planes to count his reindeer, of days down near Yellowstone before wolves were reintroduced, and how he had missed the Arctic. While his near death crash into the Chukchi Sea ended his flying career, a little accident would not affect subsistence hunting. Fred shot wolves to keep more caribou for the people of Shish. We agree that had we been born in the others' shoes in opposite geographies, we'd have reciprocally liked and disliked wolves—a détente. Different cultures, different worlds.

In 1734, Ivan Kirilov, an officer of the newly created Saint Petersburg Academy of Sciences under Peter the Great, approached a new world as he laid out goals for the Second Kamchatka Expedition. This would be one of several Russian, European, and American explorations that would have an unassailable effect on native cultures. Kirilov's sailors were "to reach from Kamchatka the very shores of America at some unknown place about 45 degrees longitude; to search for new lands and islands not yet conquered and to bring them under subjection; to write a history of the old and the new as well as natural history and other matters." The naval commander was to be Vitus Bering. The expedition's physician and chief naturalist would be a fellow "passionately in love with science, tough and indefatigable," named Georg Steller.

In 1741, Steller would become the first European to set foot on what would later be Alaska. The expedition confirmed a continent east of Asia, one not connected by land. The story has many volumes of death and intrigue, fearless animals, and warfare. Among vertebrates new to Steller, five now carry the Steller adjective—eider, jay, sea eagle, sea cow, and sea lion.

A half century later, Thomas Jefferson would also deliberate on continental connections, expressing interest in fossil elephants and geography, and pondering people and language. In 1794, he had ethnographic ambitions: "I endeavor to collect all the vocabularies I can of American Indians, as those of Asia, persuaded that if they ever had a common parentage, it will be in their languages."

Jefferson's interest in indigenous languages did not exactly inspire the mid-twentieth-century American educational system that I experienced in Los Angeles and that Fred experienced farther north in an Alaska high school. In the 1950s, Fred and other native schoolchildren did not even learn their parents' language in school because Inupiaq was not taught. English was. Native villages in Alaska often had little in the way of any secondary education. Politician and writer Will Hensley described his boyhood around Kotz during the 1950s: "My family was traditional. We were learning as our forefathers had learned: by surviving in one of the most hostile climates on earth. Most of my relatives, if they could read and write at all, did so at an elementary level.... No one ever mentioned school to me." Freddy did go to school, but that didn't come about easily.

Like other small Eskimos villages, Shish, Kivalina, and Noatak also had no high schools then. Youngsters were forced to relocate. Kotzebue in the 1950s had only a religious school. So, Freddy Jr. reluctantly attended boarding school for indigenous Alaskans in Sitka, twelve hundred miles away. He was not alone. The practice of uprooting native families to teach "America's" ways began in 1881. Nearly seventy schools across the country housed four thousand students. Freddy embraced the opportunity, but there was also a cost. "I grew up faster. I met other Eskimos and Indians. It was good, I learned a lot. It was strict and there were lots of rules. We didn't have rules growing up, not like these. I missed my family."

Species, almost by definition, solve similar problems differently. Sometimes the solutions are the same. Locale is not regularly relevant—whether desert or mountain, grassland, or thorn scrub. Long slimy tongues and strongly arched claws dominate those that eat termites and ants, earthworms, and spiders. Those consumers come from Australia, Asia, South America, and Africa. Better known as giant anteaters, echidnas, aardvarks, and pangolins, this group diverged millions of years before the ancient elephants left their tracks on the Arabian Peninsula. The convergence of form needed to suck in their insectivorous bait makes it easy to assign adaptive value.

It should be the same for seasonal variation in color. Hares from snowy environments change seasonally to camouflage with their seasonally changing background. So do some weasels and the Arctic fox. Similarity in this tactic is shared despite distant ancestry, and this is called convergence.

Color patterns of muskoxen and mountain goats are more perplexing. Both are members of the subgroup Caprinae and both live in mountains or high latitudes that are cold. Yet white goats and dark muskoxen wear strikingly different colored coats from one another. Only one of the animals is starkly obvious when the snow either melts or falls, given that neither species changes coloration.

The color of muskoxen was noted more than a century ago by Alfred Russel Wallace: "We have the thoroughly arctic animal, the musk-sheep, which is brown and conspicuous; but this animal is gregarious and its safety depends on its association in small herds. It is, therefore, of more importance for it to be able to recognise its kind at a distance than to be concealed from its enemies, against which it can well protect itself so long as it keeps together in a compact body."

I don't know why muskoxen are dark and mountain goats white. The riddle points toward a question about evolution. Conservation biologists pursue the consequences of evolution in attempts to ensure the perpetuation of biodiversity.

Wallace had been explicit that the safety of "musk-sheep" depends not on its coloration but on its "gregarious" nature, by which I assume he meant group defense. We know this is not always the case, because herds break apart, individuals run, and predation is greater during flight. While wolves have long been assumed a major nemesis, insights over the last three decades are showing that grizzly bears can also play an important role.

The reconstituted muskox population on Alaska's North Slope, farther north of our study areas, proliferated from the 1969 and 1970 reintroduction of sixty-four animals into the Arctic National Wildlife Refuge. This Maine-size, roadless, and people-less reserve once carried a muskox population of more than 425. A dramatic decline spanning twenty years resulted in less than half a dozen remaining individuals. Predation by grizzly bears accounted for some of the mortality. Of forty-six known kills before 2001, twenty-eight were by just a few bears, but twenty-two of the predation events occurred in the last three of those years. Wolves accounted for none, though monitoring in remote realms like this is always difficult. Assuming all the sampling was the same—which probably it wasn't—predation by grizzly bears increased some twenty-five-fold.

Grizzly bears cannot be the entire story because not all muskoxen perished, and polar bears had little, if any, role. Some muskoxen spread east and others west, leaving the reserve and moving into Canada or west of the Dalton Highway, which connects Fairbanks to the gas fields at Prudhoe Bay. Increased movement of a herd can signal an inadequate supply of food, a shift to dodge predators, or perhaps that subordinate animals are just dispersing to avoid the more dominant. Not knowing why the decline at Arctic National Wildlife

Refuge occurred is one of the chief reasons I focus on growth rates of individuals—in essence, to offer a more specific test of food as a limiting factor for muskoxen.

Perhaps the situation with bear predation in the Arctic Refuge was a mere anomaly, much like the curious situation at Chobe in Botswana, where lions learned to be extraordinarily bold and became adept at killing elephants. Or just maybe, something else was going on north of the Brooks Range. There had been, after all, verified reports of bears killing muskoxen far to the east in Canada's sprawling Thelon Wildlife Sanctuary.

Like much in ecology, the deeper we look, the more complex the reality. Examples of destabilized relationships with societal outcomes abound. Consider the Yellowstone ecosystem. White-tailed jackrabbits were reduced or extirpated in and around the Tetons, and coyotes altered their food habits and accounted for 80 percent of the mortality of pronghorn fawns. In areas like the Henry Mountains of Utah, where coyotes are heavily harvested, jackrabbits now consume nearly 50 percent of the grass that ranchers want for their cattle. Stepping back further—more than a century, to 1892—more than forty thousand hares were killed in a two-month period in California's Fresno County because they eat vegetation that sheep and cattle also consume. Twenty years later, in Kern County, similar slaughters occurred. Soon thereafter, residents in the town of Bakersfield called for the creation of a tax-funded mosquito abatement program. Perhaps, with far fewer hares, mosquitoes then attacked humans with a greater ferocity. The point is that unintended consequences, many indirect, occur when we mess with natural systems.

Muskoxen have a long, deep history of persistence punctuated by range contractions during periods of weather adversity—and all well before humans colonized northern Canada and Greenland. Studies of ancient genetic diversity reveal strong population contractions. During more recent archaeological periods, remoteness seems to have been the best defense, at least against humans, because Canada's high Arctic islands like Devon, Melville, and Axel Heiberg have generally supported relatively denser populations. However, even in the absence of people, muskoxen have not always had it easy. And, they've not been quick in colonizing some areas, such as Baffin Island, despite suitable habitat.

Do extremes in weather set a stage for a roller coaster in population persistence, or do humans drop muskox populations so low that rebounds are prevented by other events? What is the current role of humans where agencies limit harvest to manage populations in such a way that will promote long-term persistence? I seek Freddy Goodhope for an Alaskan perspective.

"Who really knows if they'll be here one hundred years from now?" he says.

The climate conditions of today on both sides of southern Greenland are too unstable for the country to maintain a very great population [of muskoxen]... animals need a constant Arctic climate without periods of thaw and melting snow in winter.

CHRISTIAN VIBE, *Arctic Animals in Relation to Climatic Fluctuations*, 1967

5 Muskoxen in Ice

Valentine's Day, 2011. Marci Johnson boards a Super Cub in Kotz. It's a flight that will prove fateful.

February is dark and cold—the average high is 8°F. The morning sun doesn't crest the horizon until after eleven. Marci will fly to the north end of the Bering Land Bridge, only a thirty-mile hop over the ice-clad Chukchi Sea. She hopes to find muskoxen at Cape Espenberg.

The Arctic's marine environment can be frightening. Storms are severe. Hurricane force winds fracture ice-bound seas. A rescue can be days or weeks away.

The Seward Peninsula on which Cape Espenberg is located has long, flat beaches that crawl into the shallow ocean. On the lee side are frozen lakes and ponds. Freddy Goodhope's ancestors sunk subterranean homes into sand dunes. The promontory is where he and I usually find muskoxen.

Marci zooms in low to photograph a group of about fifty at a brackish lagoon. The photos will become evidence.

A few weeks later Marci books another recon flight. The count is similar, with a total of fifty-five. The same four radio-collared females from earlier, all still alive, are with the herd yet.

Two days later, the weather collapses. After a two-week hiatus, Marci returns to the sky. She sees no muskoxen at Espenberg. Radio pulses are rapid. This means one thing: death. The light isn't good. The plane can't land.

Two more days pass. I fly down with her, hoping to spot muskoxen near the lagoon. Jagged ice covers it. Chunks of plate-like glass spread like confetti for more than a mile. We find a semi-smooth swath of ice. The plane's skis bounce across snowdrifts.

In eerie silence, strands of hair flutter on serrate ice. Blocks of crumbled ice cover sandy tussocks. The dead are barely visible. One body pokes through, and another is twenty yards farther. The carcasses are untouched, with no evidence of raven or foxes. Rings of ice have refrozen where once the bodies were warm. A bull stares expressionless. His eyes are intact, and head and horns poke above the death chamber. His powerful legs were unable to propel him beyond the ice trap. At a place where the marine environment and land collide, a spot few people on earth will ever know, we find thirty-two muskoxen entombed.

In mid-March, Marci and I draft a memo to park service supervisors titled "Muskoxen Mass Mortality Event." In it, we describe the graveyard frozen in time and underscore uncertainty, because we don't know what happened, whether any escaped, or if others are submerged. Maybe some from the group that had numbered more than fifty are still deeper in ice.

To find out, I contact the Bureau of Ocean and Energy Management in Washington, DC, for ground penetrating radar, and then ask Alaska's Air National Guard with their huge helicopters, if they would evacuate some of the dead. Neither is impressed. Maybe calls from a lunatic in Kotzebue, Alaska, who asks for help about some catastrophic ice event that he can't quite describe, and that apparently killed an unknown number of animals, is a reason to do nothing. Had they agreed, the curse of the Arctic would remain—logistics. A team would be needed on short notice to extricate and transport the iced beasts, and we'd need a heated warehouse to store and thaw animals. And, always, there is the bottom line: who pays?

By June, most sea ice is gone, a loss matched by the appearance of thousands of wriggling tawny shapes. Bleating neonates follow a milk train—caribou mothers, all plowing forth across permafrost. Most births are within a remarkable ten days, part of an evolved strategy that overwhelms the predatory prowess of wolves and bears. How caribou mothers achieve such timing is a mystery. Among bison cows, synchrony is achieved because, when heavily pregnant, they smell the vulvas of others to determine their states and then accordingly adjust the timing for when they give birth.

Late spring also means a shift from winter's white to a land of vibrancy. Moths and butterflies search for golden nectar. Tiny purple saxifrage breaks the green mantle. Cotton grasses bloom in white flowers. Chirps fill the air as hatchling bar-tailed godwits, terns, and pectoral sandpipers signal hunger. The pups of polar foxes and red foxes jostle outside dens. Daylight intensifies, growing longer every day. February's bleakness is forgiven.

June is not all bliss. The permafrost is sodden, and rivers dangerously swift. Walking any distance with a full pack is near impossible. With over a million lakes and ponds, the Alaskan Arctic is waterlogged. While the buffy white coats of Dall sheep might dot green slopes, the best way to see wildlife and the land is not by zooming a hundred miles per hour above it. One cannot feel the melting of snow or the warmth of the sun. One cannot see wolf tracks in mud or sense the thrill of a grizzly bear at close range. There is no wind in one's face, and no buzzing of bees.

From that very first instant we put into the water, I knew rafts were the way to go. Ten years earlier I had floated the Hulahula River with colleagues from Wildlife Conservation Society, navigating the river from its origin up in the Brooks Range to the Beaufort Sea. That was in 2001, when muskoxen were still to be seen on the coastal plain in the Arctic National Wildlife Refuge. Now, hundreds of miles to the west, but still far north in Alaska, I'm on another weeklong float; this time it's on the Utukok River down to the Chukchi Sea. The only muskoxen I see are on the flight north, up in the Noatak drainage some forty miles out of Kotzebue. Other herds may be to the north or west but we know neither the extent of such herds nor their whereabouts.

The Utukok highlands have Alaska's largest caribou herd, at more than a third of a million. Wolverines here are at their highest densities in North America. Whether it's because the caribou are calving is not clear. We know little about processes in this Indiana-sized parcel that lacks roads and people. By contrast, the furthest spot from a paved road in the lower forty-eight is twenty-nine miles in Yellowstone's southeast corner. Here, the Utukok ecosystem is not a vignette of a former era. It's still intact, big and wild.

The area we raft has the unfortunate name of the National Petroleum Reserve in Alaska. Its bountiful wildlife—from a multitude of fish and a dizzying array of avian migrants to the terrestrial mainstay, caribou—is a subsistence hub for the predaceous, which includes seven Arctic villages. These highlands are America's land, just like all public lands. They differ, however, in one critical way: how they're managed.

The National Petroleum Reserve in Alaska has colossal gas and coal reserves. It's estimated that 3.5 trillion tons of coal and five times that in cubic feet of natural gas were set aside there by President Warren Harding in 1923. My WCS colleagues and I have assembled here to understand potential impacts of developing these lands. What of current and impending additional seismic operations, both on land and in sea? Will helicopters, sling load operations, and heavy equipment disrupt wildlife? Are roads and pipelines here really the answer to America's energy thirst? Will coastal export operations affect beluga and bowhead migrations? Answers are absent. Local villages are split in opinion. Concern is high.

Our rafts round a bend. The shorelines vacillate between openness with deeply incised banks to gentle slopes. Willows, short and tall, line the shoreline. Two caribou splash through drenched tundra. They had not seen us; their worries must be elsewhere. I scan behind them, trying to ascertain the cause. Maybe it's a bee sting or mosquitoes, or a wolf or bear. I see nothing.

We pass a gravel bar with an inanimate ocher shape. As we near, it takes the form of a newborn caribou calf. We speculate that perhaps the mom was performing the equivalent of the plover's broken-wing display, feigning injury or, more likely, creating a distraction to lure predators away from its "nest." It all made sense, a good tale. Plausibility is different than proof, though, just as fact differs from fiction. Kipling had a name for this—"just so stories." Evolutionary biologist Steve J. Gould referenced Kipling's descriptions of how animals acquired particular characteristics when he suggested that biologists likewise use just-so stories to spin yarns about putatively adaptive behavior when proof is lacking.

We clear the spot where a caribou mother's prints depart the island. A few mosquitoes dive bomb, and we flush two ducks. Someone looks back and watches a pair of wolves exit the willows. Perhaps we were right. A mother was leading wolves away from the island.

During the next few days it's as if we see more grizzly bears than caribou. We glide silently past one napping on a mattress of winter's remnant snow. A couple of unhappy wolverines flee. There are gyrfalcons, golden eagles, and willow and rock ptarmigans. There are robins, a species that has come north only recently with expanding vegetation and insects. We float by an old fishing cabin, its nails hammered from the inside out. Mimicking the thorns of acacia that deter giraffes bent on their fleshly content, the local Alaskans hammer spikes to deter bears from break-ins—convergence in deterrence. In this case it's between plant and human to thwart hungry animals from their nutritious wares. Similarities in form and function of course, but what interests me is that over in Asia the Chukchi people also use nails in their rustic abodes to reduce similar behavior by bears. Their bears are white. This is cultural convergence in response to grizzlies and polar bears.

Not all polar bears retreat with the sea ice northward, and some remain on land. Others have drowned including at least one mother and young trying to swim more than five hundred miles to the nearest northern ice. And others just get shot. One of the unlucky bears encountered a hunter some two hundred seventy miles inland from the Beaufort Sea, and died just a bit north of Fairbanks.

We float closer to the coast where the embankments collapse and expose melting permafrost. Instead of land where glaciers carved the topography into couloirs and U-shaped valleys, the terrain alongside this section of river grows gentler. It's where the hooves of saigas and camels once crossed turf, a place where native horses grazed alongside giant bison. Ice glistens in a late evening sun. The meandering Utukok discloses chunks of ice in patches of vertical soil.

Perhaps mammoths anointed the very ice we now collect. Conservation biologist Kent Redford needs it to mix with his newest gin-based beverage. I chip Pleistocene ice. He names the brew the Pleisto-cini in honor of the martini. We drink too many, grow silly, and speculate that had pachyderms trodden above our ice reservoir, time would have long since erased the toxic elixir of gallons of elephantine pee—a mix of urea, sodium, and potassium. Geeks, for sure.

Our pickup flight will take us to Point Hope and then to Barrow. We fly above gentle hills. Labyrinths of matted vegetation coalesce, lined up in a single direction. Just as tracks in sand and snow tell stories, these sodden paths hide some.

We cross over more ridges and more padded tussocks. Routes crisscross, disappear in water and reappear. Suddenly, we're over a colorful swarm—thousands of yellowish-brown objects. A single mass moves and splashes, babies streaming after adults. We've caught the tail end of a hundred thousand migrating caribou. They're heading to the coast where it's windier and mosquitoes fewer. June is warmer now than it used to be. The mosquitoes are earlier. More change will come. With warmth there will be more melting, more mosquitoes, and less ice.

Death assemblages are uncommon, especially for muskoxen. Outside the Arctic, such mass mortalities are not. In the 1830s, Charles Darwin described a mixed accumulation of dead ancestral wild camels and horses. They perished only a few million years ago on South American grasslands. He didn't know why.

Around then, but some sixty-five hundred miles to the northwest in what is now Idaho, a volcanic explosion spewed ash. It reached beyond Antelope County in Nebraska. Fast-forward twelve million years to 1971, when geologist Michael Voorhies stumbled across a few thick bones in a dry streambed. He and his team would later report a dozen barrel-bodied rhinos that had suffocated suddenly from the toxic ash.

Animals confront natural hazards all the time. Rarely do we know much about the magnitude of the threat, its frequency, or the fates of nearby animals. Many go missing, like our vanished twenty or so muskoxen. A one-time kill event may be considered a pure anecdote, but when anecdotes are strung together, the stories and questions get more interesting. Are the animals just the unlucky or do they have traits that predispose them to additional mortality? How deadly are natural events and how do they affect the population structure?

One day in November 2002, biologist Alberto Scorolli was out hiking the grassy hills in the pampas of his native Argentina. The feral horses of Parque Provincial Ernesto Tornquist are Alberto's passion. Summer in that part of Argentina is warm and sometimes wet. On November 9, the rain began. By the time forty-eight hours had passed, ten inches had fallen. At the storm's onset the temperature was 80°F. The next day—still raining—the high reached 45°F. Lows dipped into the thirties.

Horses tend to be hardy, but these subtropical ones were of a different nature. Sopping wet and chilled by winds that blew at thirty to fifty miles per hour for thirty-six hours, more than 190 died, which was 25 percent of the entire population. Alberto's findings, described in "Unusual Mass Mortality of Feral Horses during a Violent Rainstorm," are notable because of the extreme mortality rendered by a single subtropical rainstorm.

Unusual weather-related deaths occur at all times of the year, but events logically vary by continent and species. Lightning has killed entire cattle herds with a single strike in Uruguay and Zimbabwe; one killed some three hundred wild reindeer in Norway. In Montana, five hundred pronghorn perished during heavy snows. More surprising, perhaps, is the occurrence of rare snowfalls in South Africa that killed entire bands of endangered mountain zebras. Droughts are obviously deadly to a wide array of species.

Other sorts of unpredictable events occur as well. During Yellowstone's 1988 fires, a hundred fifty elk died. The 1980 Mount Saint Helens volcano killed twenty-six. Canada's eastern boreal forest mired fifteen moose in a single sump. And two similar events claimed the lives of the so-called New World's largest land mammal, the bison. Both of these occurrences involved more than two thousand animals crossing flooded rivers, one on the Platte River in Nebraska during the nineteenth century, and the other late in the twentieth century in northern Canada's Peace River.

In a situation vaguely reminiscent of Darwin's discovery of fossil camels and horses, my fellow travelers and I once stumbled upon a stench-ridden bog among sand dunes in the Gobi Desert. In varying stage of decay were thirty-eight two-humped camels, wild asses, and gazelles. In these sorts of events, causes of mortality can often be assigned. Rivers and bogs are sometimes predictable in time and space, but volcanoes and fire less so.

The muskoxen death site north of the Arctic Circle was frustrating because we had no idea about the cause. We also didn't know the number of dead, or whether we'd return before the spring thaw and before the bodies were lost. The types of events that trap and kill differ, even though a barrage of severe storms, volcanoes, and earthquakes all have an impact on animals. Earthquakes bring tsunamis, and volcanoes bring ash. All foment destruction. So does extreme snowfall.

In September 1992, an unusual four-day storm dropped more than three feet of snow in Denali National Park, over half the yearly total. As a consequence, 75 percent of the radio-collared caribou moved more than a hundred thirty miles over the next few weeks. The normally sedentary park animals found less snowy ranges, some beyond Fairbanks.

When weather events cross immense geographies, their causes and effects on wildlife are not easily discernible. While the long-distance movements of Denali caribou in the winter of 1992–93 likely prevented significant mortality, an occurrence halfway around the world could have been the loud trigger. In the obscure Cabusilan Mountains in the Philippines on the island of Luzon is the now infamous mountain called Pinatubo. On June 15, 1991, it erupted, with the second largest explosion of the twentieth century. Its noxious aerosols formed a layer of sulfuric acid that cooled the entire globe during the next year. Whether a high-pressure system north of Russia's Wrangel Island was connected to the Pinatubo explosion is not clear. But, that very storm north of the Chukotka coast brought the unseasonal cooling and early heavy snow to Denali.

Natural hazards will always affect animals and people. Sea-level rise is continuing to assail Arctic coastal villages, perhaps even lagoons like the one with the entombed muskoxen. Warming temperatures will reset hunting seasons and migration, just as they affect the timing of marine and river travel by subsistence hunters, because the duration of navigable ice lessens.

Storm surges are also destructive. In 2012, Hurricane Sandy caused $50 billion in damage; six hundred fifty thousand homes in New York and New Jersey were among those affected. Embankments and barrier islands failed against twelve-foot waves that pummeled properties and moved inland.

Alaskan natives have long appreciated the value of berms; among the most treasured is ice, which at times plays the role of barrier islands. In today's world, with the later freeze-up of marine waters, villages are increasingly vulnerable to coastal flooding and erosion. Wind-propelled storms generate bigger waves, and the forces are even stronger when broad stretches of water remain unfrozen. Fetch, as the meteorological phenomenon is known, is dampened when sea ice is grounded near shore. Called landfast ice, it reduces the impacts of surging waves and water on land, especially on the vulnerable low-lying regions along coastal Arctic Russia and North America. Offshore events have onshore consequence. Nature connects us all.

In spring of 2001, Freddy Goodhope was up at Cape Espenberg. The breeze was light. Corrals that once confined his reindeer no longer confined any living thing within them. He was uneasy, as if being watched. Freddy rarely felt like this. He saw movement far off but wasn't sure if the something was covering distance. With no binoculars to detect what it was, he just waited. The something was squat and not caribou. It was moving toward him. With little vegetation to compare with, judging size is tough. Sometimes porcupines look like grizzly bears at three hundred yards.

As the animal ambled, Freddy could tell it was not brown like a grizzly. It moved sideways, then directly toward him, then parallel. This was not to be an accidental encounter.

At sixty yards Freddy fired a warning shot. The polar bear stopped. It stared. Its paws were not huge. It sniffed. Its ears were back, but then erect, almost curious, nose high. It kept coming. A second warning shot. As it drew closer, Freddy aimed, and squeezed. The white bear moved forth a few steps and dropped.

The young male was in good shape, not starving, simply curious. Curiosity can kill, just as it did young gazelles in Serengeti. Maybe the polar bear had never seen a human. Bears do not have happy outcomes when combined with people.

One unpeopled place some five hundred miles south of the Espenberg Peninsula and two hundred miles west of Nunivak Island is

Saint Matthew Island. This very secluded place rises fourteen hundred feet and covers almost a hundred fifty square miles in the North Bering Sea. Cold and windy, it's occasionally surrounded by firm ice. There are no muskoxen. There are no people. Rarely are there polar bears—but this wasn't always the case: polar bears once summered there. Knowing that helps explain the changes we see today—those caused both by altered climes and by people.

In 1809, the Russian-American Company attempted a small settlement under the guidance of Demid Kulikalov. It didn't work out. Saint Matthew and the surrounding, smaller islands languished in obscurity. Sixty-six years later, they would become known colloquially as the Bear Islands because their white denizens abounded on the rocky scarp. We know this because as nineteenth-century whalers tracked their cetacean quarry, they recorded what they saw as they followed the retreating winter pack ice.

A drawing of polar bears covered the 1875 frontispiece of *Harper's Weekly*. Inside, H. W. Elliott wrote: "We landed on [Saint Matthew] . . . early on a cold gray August morning, and judge our astonishment at finding hundreds of large polar bears . . . lazily sleeping in grassy hollows, or digging up grass and other roots, browsing like hogs." In 1876, the expedition's lieutenant wrote: "The females were accompanied by their cubs, from one to three in number, which were (in August) about one third grown." In September of 1885, a period when the winter ice had not yet formed, at least four polar bears on Saint Matthew Island could be seen from a ship, the supposition being yet again that there was an indeterminate number of summering individuals ashore. David Klein and Art Sowls concluded their 2011 investigation of polar bears saying that

> the last record of summer-resident polar bears on the St. Matthew Islands appears in a brief paper on the mammals of St. Matthew Island by G.D. Hanna (1920), who noted that crews from the Revenue Cutter *Corwin* killed 16 polar bears there sometime in the 1890s.

> In 1899 the Harriman Expedition, which included the eminent mammalogist C. Hart Merriam and the pioneering conservationist John Muir, visited St. Matthew, Hall, and Pinnacle islands. Muir and Merriam were particularly chagrined to find no live polar bears.... They did report finding skeletal remains of bears with bullet holes in the skulls.

In 1972, the possibility was raised of reestablishing a summer population on Saint Matthew. It did not garner steam. Now—approaching fifty years after that initial suggestion by the state of Alaska—ice continues to decline not only around Saint Matthew but well into the Arctic. Polar bears need ice platforms to hunt their primary prey—seals. Where they've been losing ground is where ice disappears. At the bears' southern terminus in interior North America, a place called Churchill at the bottom of Hudson's Bay, their fecundity and body mass have decreased over two decades. Polar bears would not be able to recolonize Saint Mathew. Ice recession is reshaping the Arctic environment. Its denizens are not immune. Again, offshore events have consequence on land.

It was an hour after sunrise as we flew south from Kotzebue toward the Bering Land Bridge in March 2011. The thirty-mile stretch across the Chukchi was no longer thick with ice. Pressure ridges that had once been sutured were splayed apart. Open water and long leads reflected a horizontal sun. Flotillas of hexagonal ice drifted, with a few bergs the size of ships. Our flight time would double as a consequence, not a trivial expenditure when helicopters cost more than a $1000 per hour. Federal safety rules were in play, however, and require following the shoreline when water is open. We used the opportunity to scan west and south, hoping to sight polar bears, what Norwegians call *isbjørn*.

Polar bears are not usually abundant in my study regions. That's why Freddy Goodhope was so surprised to see one at Cape Espenberg. Every now and then one comes into Shishmaref or Kivalina. One had crossed through the Igichuk Hills, its path only a couple of miles from muskoxen. A local fellow on a snowmobile shot it, suggesting that he feared for his life. The story raised eyebrows since polar bears are not happy about the revving of snowgos and probably would have run off. Another bear had wandered into the town's outskirts and bluffs east of Kotzebue once where I had been out jogging. I happily learned this only afterwards. That bear also disappeared.

In rural communities, there is a local vernacular for unwanted varmints who vanish. For wolves in the lower United States, it's called the triple *S*—shoot, shovel, shut-up. Up here in the Arctic, the second *S* won't work since there's an impediment: permafrost. One can't dig, and so the carcasses are there for discovery.

Our helicopter flight continues. As we peer onto shorefast ice, the leads grow longer, some a mile or more. Seals duck into the water, sending circular ripples. We turn right. In another lead are ringed seals; another lead has more seals, and then more. A few bearded seals are among them, and a quick good glimpse confirms a spotted one. We hadn't expected the Serengeti of ice with seals everywhere, by the thousands. If we have seals, shouldn't there be bears?

We chug along, looking down. Maybe we'll find a spotted seal four-miles inland, as had others, or see a bearded one scurrying along after its ice hole has frozen solid, to then be bloodied by Arctic foxes. No such luck.

And then, all of sudden, there it is: a white fox is running, jumping.

No, not a fox—what we're seeing is big and furred. And though it's not lumbering, it's a bear, which is jumping small pressure ridges, intent on open water. It dives, swimming to the next ice floe.

A few miles of solid land pass and then an odd stain comes into sight. Circling closer for a better look, a pink insignia is squiggled across snow, as if painted by a child Rembrandt. Confused, we backtrack and discover a deep hole in the shore ice. The pink stain grows to a deeper red. A huge half-eaten seal still drips blood. Two ravens are on it. So is a polar bear; it's not running, and we pull away. Over the next week we see more seals and more polar bears; these include mothers with cubs, mothers with larger cubs, and solitaries.

What we are observing is atypical of this area, although polar bears occur across all of northern and northwest Alaska, depending on ice conditions and consequent seal abundance. Canada has the greatest number of white bears, but population segments also come from Russia, Norway, and Greenland. The long leads and dismantled pressure ridges of ice north of Espenberg have created the display below. I wonder about changing regimes of snow, ice, and water and what, if anything, this has to do with the entombed muskoxen.

The dead muskoxen in ice nag. I'm still clueless about the fates of the missing twenty or so. One fine spring day with temperatures surging to ten above, Freddy, Wibke, and I click off sixty miles on our snow-gos to view the iced graves. Fred's never seen anything like this, not with muskoxen, not with caribou, not with walrus.

Then the winds whip up to fifty miles an hour, and we find ourselves stuck there. To go far from Fred's Espenberg cabin to pee becomes challenging, as the visibility dips below twenty yards. Our snow machines are buried. We have days to wait out the storm. We discuss death assemblages.

A few possibilities are quickly discarded. Starvation is out because the animals that Layne and I handled from other areas were in poorer condition and with less body fat than were the dead ones, yet the hungry lived. Further, the current winter has not been unusually cold, and the prior summer was no different in temperature from the past ones. Most animals should have been in decent shape heading into winter. Similarly, predation was rejected as a cause. Although both grizzly bears and wolves exhibit a rare behavior called "surplus killing," whereby more animals are slain than can be eaten, we had yet to see any evidence of bears or wolves. Also, no carcasses had been fed upon. Only a curious red fox had visited. No ravens were evident. There were no indications of a human role.

Another alternative, disease, might have been operative, especially since muskoxen carry chlamydia, brucellosis, parainfluenza, and others. Unlike lightning, however, disease rarely takes out all animals at once. Floods, though, do.

And so we surmise that our animals died either from drowning or hypothermia. The bodies held answers. So might the twenty or so as yet unaccounted for. If alive, they'd offer a valuable contrast.

Storms pound coastlines regularly, and in the Alaskan Arctic at least ninety serious ones battered shorelines between 1900 and 1980. Among the largest occurred in 1963, when the still active Naval Arctic Research Lab operated in Barrow. The wind had blown with a fetch of more than three hundred miles in the ice-free Beaufort; Barrow was slammed for three days. Ten-foot-high waves pushed water twelve hundred yards inland. The lab was flooded, as was town. Sea ice piled up more than thirteen feet high, bluffs lost nine feet, and shorelines eroded about twenty yards.

Storm intensities are now increasing throughout the Arctic. To the south of Nome, on the Yukon-Kuskokwim Delta, massive gales hit in 2005, 2006, and 2011, and flooding swept up to twenty miles inland. Hundreds of miles to the east in Canada's Mackenzie Delta, a high-energy event in 2001 pushed salt water a dozen miles into the interior. Twenty square miles were flooded, killing vegetation and fish in interior lakes. Wind and salt water bring in a nasty amalgam of gravel, sand, clay, and silt—an admixture unkind to land-loving plants and animals.

Knowing that offshore events affect onshore geology, hydrology, and vegetation, Jim Lawler and I investigated databases to gauge the magnitude of the particular storm that pummeled Espenberg and its muskoxen. We checked National Oceanic and Atmospheric Administration databases between Barrow and Nome. A picture began to emerge, one that included the Red Dog Dock site on the coast between Kivalina and Kotzebue. This is the only location along Alaska's western Arctic coast where tides are measured. From these Red Dog measures, we developed an idea of what may have occurred at the Espenberg lagoon during that February stormy period.

The normal difference between high and low tides in February is maybe eight inches. Between February 18 and 26, surges reached more than sixteen times that, and that's when something happened to our coastal-dwelling muskoxen.

Ice shelves near land are normally attacked by offshore ice. Consistent friction and slamming, retraction and re-ramming between old ice and new, or ice and big waves, produce buckling. Pressure ridges can soar to twenty feet. In deep water, pressure ridges can anchor ninety feet below the ice's surface. In shallow seas and lagoons, of course, this cannot be the case. At the Espenberg death site, ice plates were pushed up and shattered on land, some beyond sandy hillocks. Soil and vegetation that normally is under snow or ice was torn loose, thrust forth, and reburied more than three hundred yards from the lagoon's edge. Large numbers of long dead tomcod and herring were discovered up to a half mile inland that summer. Ice tsunamis are what some call these storms.

More appropriately, the events are ice overrides, a form of marine subsidy that brings frozen seawater to land. This happens when sea ice is propelled ashore by furious winds, like the February 2011 storm with its gale-force winds of sixty to hundred miles per hour.

Brutal, loud, and of unfathomable power, ice overrides are rooted in traditional ecological knowledge. Called *ivu* or *ivuniq*, the force from this polar-based tidal surge is said, in the *Iñupiat Eskimo Dictionary*, "to compress, to form a pressure ridge in ice." Winton Weyapuk from the small village of Wales (the closest point to Russia, only twenty-eight miles away) said of an *ivu*'s power: "The pressure ridges look as if they have been sheared off and a new row of ridges formed closer to shore. . . . Now the *ivu* . . . comes in fast and gets shoved up on the beach when the wind is strong."

What must have occurred at Espenberg was that massive, fractured ice and wells of water engulfed the luckless animals. Some tried to make it to shore but were immersed. One male perished while standing frozen in ice, head up, chest firm, legs planted. Some may have died from hypothermia. Drowning is far more likely. Had their kidneys or lungs been accessible and had we been forensic investigators, we'd have searched for bacterioplankton, a sign of drowning. They weren't, and we weren't.

When death occurs from drowning, it's not because of water in the lungs. It's an oxygen deficiency to the brain and heart. The larynx can shut down, but the lungs may be have no water in them. To know the true cause of death, we needed to necropsy bodies.

By May the ice was still thick. Layne, Marci, and Eric the pilot flew back to the lagoon with power tools; the most impressive was a saw to cut through blocks of ice and remove the animals' heads. Every animal was mapped. The missing twenty were now found: all were in ice. The total now numbered fifty-two, which meant that, of the original fifty-five we had photographed in February, three had escaped. Of the corpses, three had part of their horns sheared off in the *ivu*. We took the heads of a couple of others and flew them, in a wretched but frozen condition, to Kotzebue and then on to the Museum of the North in Fairbanks. Globs of ice were still lodged in their pharynxes, a clear indication that they'd drowned.

Among the dead were twenty-five juveniles, including eleven yearlings and fourteen two-year-olds. The rest were adults. The inlet itself claimed 72 percent of the bodies. Thirteen others were submerged in ice on land. If the herd was clumped at the time of the *ivu*, as the photos prior to the event showed, then some animals fled and made it to shore. Of course, we didn't know how the animals were actually arrayed when the storm hit. The body furthest from the inlet's edge was 375 yards out toward sea, and the greatest distance between that and another animal was more than a mile inland. The most intense assemblage of ice-encased corpses, twenty-seven in all, was an area of the inlet about 275 yards wide and 125 across.

Two months after the fifty-two excavations of muskoxen, Freddy was back up near his cabin. He watched fifteen different grizzly bears in a single day heading to the bloated muskoxen bodies. A summer archeological camp with high school students was also in the region, and because safety is paramount, extra vigilance was called for.

Archeologists visiting Espenberg weren't especially worried about bears or about taking precautions. "Damn muskoxen," one lamented, "they trample our sites." The issue for them was not about the recent dead or the *ivu*, it was about enigmas of Beringia—how earlier peoples lived. Muskoxen obviously complicated the archeologists' scientific sleuthing. For me, however, the pursuit is about change on land and in sea, how wind, water, temperature, and snow affect the Arctic's largest terrestrial mammal. Just a few months before the archeology camp, a furious and unprecedented weather event iced a herd. Many secrets remain, among them the frequency of similar events that go undetected worldwide.

> We saw rain in November and December. Usually, our fishing starts at the end of October, when the lagoon freezes over and we can get to the mainland over the ice. But this year we couldn't fish until mid-December. That means that we have to buy more Western foods–spending more money each month that we can't hunt our own food. Or we have to travel further during those times to try to hunt caribou.
>
> TONY WEYIOUANNA, testimony delivered to the Bicameral Task Force on Climate Change, Washington, DC, 2011

6 When the Snow Turns to Rain

North America's most extreme settlement is Ausuittuq—the place that never thaws. It's where the Canadian government resituated some of its indigenous population a long time ago. Seven hundred miles north of the Arctic Circle in Nunavut, Canada's newest and most northern territory, Ausuittuq's average annual temperature is two degrees, which is twenty degrees colder than Shishmaref's. Ausuittuq is also known as Grise Fiord, so-called because the grunting of walrus that explorer Otto Sverdrup heard when he visited more than a hundred years ago reminded him of pigs—*grise* being the Norwegian moniker for pigs. Subsistence is life for those who live in Ausuittuq, and local wildlife plays an important role in that subsistence. Resident Bobby Nungak told a journalist: "I love raw narwhal." Similarly, Anna Marie Qamanlg said she likes "cooked or raw caribou with salt."

If resident animals desire a future other than as frozen slabs, like the ones I shared back on the floor of that hunter's cabin in Kotzebue, they need to be wary. They also need food, mates, and shelter. Adaptability is critical. So is ice. It governs the geography of life by dictating the availability of species to be hunted.

In the late Pleistocene, a section of Nunavut was covered by what we now call the Laurentide Ice Sheet, which was nearly two miles thick. Neither animals nor people lived on it. But ice's imprint is indelible. In the wake of its glacial melting lay the Great Lakes and Lake Champlain. The scoured valleys and remnant moraines, ridges of gneiss and sand, reflect its recession, as does the resulting distribution of species sought by past and present humans.

The Ungava Peninsula of northwestern Quebec is one place left with this sort of glacial geology. This lonesome spot in the Arctic is windy, cold, dry, and therefore perfect for muskoxen. Its permafrost sustains thousands of caribou. Why no muskoxen?

The answer lies at the intersection of ice and geology, coupled with biology. The Ungava has been deglaciated from Laurentide ice for maybe sixty-five hundred years. The habitat would appear suitable for muskoxen, maybe even grizzly bears, which also are missing today. Broad similarities exist between Ungava's tundra and that northwest of Hudson's Bay, where both species exist. An experiment to test whether the Ungava is suitable for both species would be outrageously expensive. Foolhardy is another word for such a test.

But an experiment was done. In 1973, muskoxen were introduced to a patch of the Ungava for husbandry. The animals subsequently escaped or were released and now not only flourish in northern Quebec but have also spread to adjacent Labrador. Lands that once lay fallow of muskoxen were clearly suitable to them. Their lack had been an accident of history promulgated only by distance and recent melting. As for grizzly bears, a remarkable find in 1975 confirmed that the deglaciated habitat was suitable. An archaeologist discovered the skull of one from the 1700s near an Inuit sod house.

Timescales can be long or short. They shape how we consider past impacts, including those of glacial processes that began long before today's land was clear of ice. Muskoxen never occurred on Ungava until recently, although the habitats created by the loss of ice were a good match for them. If our lens is only the present snapshot, we'll forget systems are dynamic. The concept of ecological baselines bears relevance. Rather than looking across sixty-five centuries of glacial history and changing ice on the Ungava, let's look at a shorter time span in the Aleutian Range of southwestern Alaska.

Some of it is presently submerged, a land pocked with island uplifts piercing the Bering Sea and unfolding from Anchorage westward almost to Kamchatka. At the distal end of the archipelago, biologist—and ex-baseball college pitcher—Jim Estes once watched sea otter populations grow from near extirpation to local abundance. Then it all changed again. "Back in the old days . . . we probably would have seen 500 otters by now."

Ironically for Jim, "the old days" were a mere ten years earlier. A regime shift occurred in the 1990s, a tipping point, when very few otters were surviving. Following sea otters' overhunting by Russian and later American mariners, they had enjoyed buoyancy and recovered populations, but then killer whales had become their new predators. Ecological history can be set thousands of years ago, as when glaciers recede, or can occur immediately, when a new predator arrives. Time matters.

Variation in demography may be normal, or expedited by human causes or other events. Warming environments have led to glacial recession across deep time (spans of long geologic history on Earth), and the temperatures we now see at the southern terminus of species ranges already affect their survival. Mammals of high latitudes typically have trouble with heat and finding ways to minimize it. If they can't cope with it, they die out. Pikas, a close relative of bunnies and hares and a cold-adapted specialist, use tubules left behind after a lava flow for cooling when in warmer environs, like down near sea level in Oregon's Columbia River Gorge; temperature there can reach 100°F. Ungulates both here and on other continents, like ibex, chamois, and caribou, lie in snow patches in summer or find shade. When neither is available, they use windy ridges.

Caribou usually shift to higher elevations. Mountain goats, the nearest North American relative of muskoxen, also migrate to high elevations, where they use snow, caves, or shady groves for cooling.

The modern distribution of muskoxen includes a wide arc across Arctic Alaska, northern Canada, and Greenland. Mountain goats' range includes high-elevation cordilleras like the Cascades and Rocky Mountains from Washington, Idaho, and Montana, to Alberta, British Columbia, and Alaska. Like muskoxen, a good chunk of the goats' lives are spent in cold and snow. Despite the color difference, with goats being white and muskoxen dark, both species were once scattered much farther south. Native goats have since perished in Wyoming, Colorado, Oregon, and Utah. The same is true of muskoxen relatives south of the continental ice sheets that stretched to Illinois and West Virginia. Climate must have been involved, just as we presume it was in places where pikas existed and no longer do, like the scorching desert basins and mountains outside Las Vegas. Even contemporary moose now wilt where past populations boomed in Minnesota and Michigan and Wyoming. Whether species adapt to an enemy like heat and experience ecological stability depends on our temporal framings. The population segments that survive exhibit flexibility under conditions that are less heat prone and less onerous than where their southern or lower-elevation counterparts could not.

The impact of weather events, like rain during winter, are increasingly scrutinized. For instance, midwinter Arctic rains turn steely jackets on lichens and grasses once the normal cold returns. Snow surfaces become so hardened after refreezing they cannot be splintered by cloven hooves. Craters cannot be dug to access food due to the icy cover. During such periods, caribou and muskoxen consume less and spend more energy searching for nutrients. Ptarmigan, which access the sites chewed up by caribou and muskoxen, must likewise forgo access to the iced-over vegetation. Nature's web goes cold.

Winter temperatures, in particular, are under rapid assault. In 2015, Kotzebue reached a balmy midwinter high of 37°F, a new February record. The temperature did not drop below freezing that day. That was the same year President Obama visited. On January 1, 2017, the temperature again exceeded 32°F, as it did eleven months later, in mid-December. One March morning in Alaska, I woke up to a temperature of −16°F. Three days later the mercury would reach 44°F. I couldn't fathom what a sixty-degree temperature shift in winter meant to animals. For us, winter jackets that were dry in the cold became waterlogged in rain. In wetness and grueling wind, we grew hypothermic. Snow machines overheated. Thick sheens of river ice lost outer coats. When rain-on-snow events occur, like the ones in Tony Weyiouanna's testimony, a new wave of challenge arises; this is nature's rising tide of ebb and flux.

One such rain-on-snow event on Banks Island, which is eighteen hundred miles northwest of the Ungava Peninsula, famously killed twenty thousand muskoxen in 2002. The year before, the population had been seventy thousand. Despite the hardiness of muskoxen, when rain-on-snow events occur, they can be fatal, much like the single event that was fatal to Alberto Scorolli's rain-cooled Argentine horses mentioned in the previous chapter. Banks Island is remote, large, and without people, except for the small community that lives in Sachs Harbour. No one in that community had detailed the mechanisms by which the estimated twenty thousand muskoxen had died.

Not only do midwinter warm rains turn Arctic snow deadly on land, but when sudden temperature swings go in the opposite direction—turning water to ice—the mix also turns lethal for marine wildlife. More than a hundred fifty narwhals died, entrapped in ice in the eastern Canadian Arctic during a single event in the 1990s. For a hundred seventy belugas, a sudden temperature drop was similarly fatal when nine polar bears converged on a sixty-five-foot by twenty-foot opening in the ice; the bears killed one hundred of them.

Alaskan sea otters in the Bering Straits are not immune to this kind of bad luck, either. Foragers of mollusks, sea otters can sometimes be found as far as twenty-five miles from the closest shore. Ice floes dictate their northern limits in that otters try to stay south of ice floes. During the winters of 1971 and 1972, cold and wind propelled ice farther south along the Aleutians than normal. In 1971, hundreds became trapped and more than forty carcasses were located on frozen beaches. Others were nine miles inland. Biologists Karl B. Schneider and James B. Faro noted that, in 1972, "otters moved over 250 km [155 miles] in 15 to 20 days" to stay ahead of the ice pack. Snow depths reached more than two feet, and more than two hundred sea otters starved.

Extreme temperatures and unusual weather events, either at sea or on land, can be debilitating—whether to maritime otter or to Arctic pseudo-ox. Death comes in sundry ways: hypothermia, starvation, accident—and a fourth way, as well.

Grizzly bears are essentially absent from North America's high polar archipelagos, though some take northward journeys, and sightings are increasing on Arctic islands. *Canis lupus* have been there for millennia. They survive on big game. Grizzly bears do not.

As omnivores, bears subsist on vegetation, carrion, fish, and insects—even ants and moths. However, consumables like baby caribou, elk, and moose are certainly not turned away. Neither are baby muskoxen. In the early 1990s, two grizzlies encountered forty to fifty muskoxen near the Thelon River. The muskoxen, along with their eight calves, bolted. In just a mile, the Canadian permafrost had four fewer muskox babies.

Alaska's tundra is home to both grizzly bears and wolves. Prey survival is determined by what's immediately available. Caribou migrate seasonally. Muskoxen do not. If wolves follow migratory caribou, the shaggy beasts I study should face decreasing pressures. Put simply, if wolves track only caribou and with bears in winter torpor, muskoxen need only deal with vicissitudes of weather and food to survive the cold season. I was wrong on both counts.

First, consider hibernating bears. It's March 26 in 2011, and it's chilly on the tundra beyond Kotzebue. The low is −17°F, but the high rises to 7° above. We see three grizzlies—a solitary male and, later, a female and her yearling. The year before, the first bear up out of hibernation was spotted on March 8. Thus, muskoxen encounter bears during the late Arctic winter.

Second, consider wolves. I thought I was grounded in believing their packs followed caribou herds. After all, three hundred thousand migrated south and east of Cape Hope, Point Lay, and Krusenstern that same winter. The Igichuk Hills had none, nor did the Agashashok or Noatak River basins. The migratory swarms were a hundred miles away, down in the Selawik. The winter had been unusually snowy and so caribou were far south that year.

But, a wolf pack remained up near Red Dog, north of the Noatak and just beyond the boundaries of Krusenstern, and there were no caribou. Several radio-collared muskox herds were there, however. I remembered my days back in the Tetons and Yellowstone, where wolves learned to take more dangerous game including bison and moose when elk grew scarce. Why wouldn't the wolves up at Red Dog be switching prey too?

One day that year—the same year as the *ivu*—a group of fourteen muskoxen were clumped together in a hilly valley. A few days later a mortality signal pinged on one of our collared muskoxen. From a plane, we could see that cratered into deep snow was a bloodied carcass. Herd mates were resolute, refusing to move from their dead mate. Wolf tracks abounded. A couple hundred yards off, a pack rested. We still had other females to immobilize and collar, so we couldn't investigate the situation there right away.

But two days later, our helicopter follows GPS coordinates north to a secluded valley and the carcass. Standing next to it is the herd, now numbering twelve. A cluster of ravens lift off. Something is unusual; nearby, more fur and bones and blood. A second animal is dead, fifty yards from the first. The living herd still waits, patiently, now for a full three days, as if somehow their presence will usher their two dead companions back to life. There is but twenty yards separating the dead from the living. A demonstration of sentience?

I wonder about the differences among mammoths, muskoxen, and modern elephants. In Africa, elephant females attempt to revive dead and injured mates, recognizing the bones of the deceased. The group of muskoxen I watched lingered for days after wolves killed band members. Did survivors await a saintly miracle for their friends to rise? Maybe muskoxen are more socially complex than credited. Little is known of their social relationships or how their brain function changes with age.

Mammoths had big brains—about eleven pounds. Elephants are also large brained and, relative to their body, have an encephalization quotient in excess of most mammals. The encephalization quotient for modern female elephants is greater than two—twice that of males. Dolphins and primates also have large brains for their body mass; the well-endowed organs are capable of long-term memory and promoting complex lifestyles. Chimpanzees have an encephalization quotient of nearly two and a half. We're at the top, with seven.

Among ungulates, several with relatively large brains that display cognition and deep memory also live in socially complex societies. Two species of zebras, mountain and plains, live year round in bands with stable membership. Wild horses in Argentina, Nevada, and elsewhere are similar socially. They know and remember herd mates. When individuals become detached and later reunite, their recognition of each other is clear. They avoid breeding with relatives, a behavior due to past familiarity and not some sort of genetic predisposition.

Even bison, which once lived in vast comingling herds, show strong affinities, with females preferentially associating with friends and kin. We know this from detailed studies of intimate bison affiliations from herds in South Dakota.

Complexity is often incumbent in group-living societies. We hadn't understood that bonds among female baboons enhanced the survival of juveniles until we knew something about individuals across generations. We do know something about sociality in some ungulate herds, about some birds that flock, and about some fish that school. For extreme species of the north however, secrets beg discovery.

Our helicopter touches down, showering snow onto the carcass. Despite the blinding outwash, the séance continues as her herd mates seem to await a ghoulish revival. I grab my tool bag, duck under the spinning rotor, and lean down to remove her collar. A chunk of leg is gone. A hole punctures her abdomen. Part of the rump is eaten. Wolf kills are messy, lots of blood and torn tissues. She didn't fade into breathless stupor like a wildebeest throttled by a lion.

The cow lies in the snow, edges melted away by the warmth of her decaying body. I push down on her throat. Her eyes open. Mine wider. *Holy shit.*

Shock overtakes her. She thrashes, trying to stand. She can't. Layne hurries over and puts a .44 slug in her brain. She doesn't die. He shoots again, she quivers. The misery ends with a third shot. The group moves off. I check the other female, fifty yards away—cleansed by wolves and scavengers. No life there. On the backside of a hill lies another mort. Herd size down to eleven. This wolf pack needs no caribou.

Each carcass lies in significant snow, up to eighteen inches deep. Perhaps this group just couldn't escape the storms that dropped more snow inland than stubby legs can navigate.

I'm perplexed by another fact. This herd has no adult males, only a preponderance of two- and three-year-olds. Did the adult females stand helplessly or try to defend themselves and adolescents? Did snow depth hamper maneuverability? Three died. *What did I miss? Probably a lot.*

We decide to collar a female from the group of eleven since we'll need to relocate them. The darted cow goes down in fifty seconds—not good. Her sacrum and pelvis are almost pure bone. There's no body fat. After she's collared, we reverse the effects of carfentanil by giving her an antidote and before she's even up, we're back in the air, moving away. So is her group, already four hundred yards distant, plowing through deep snow to escape us.

The next day, the collared female moves several hundred yards but she's now chest deep in snow and fights once again to extricate herself. This is good and bad. At least she's alive. But she's also alone and mired. A couple of days later she's dead. Was it stress, starvation, or wolves?

A wolf pack now stands not far from the group. Once fourteen muskoxen in size, now it's seven. Custer's last stand? Deep snow, little experience with wolves perhaps, and weakened animals all conspire to challenge survival.

Death is not gentle. Over the years, we've found two-year-olds that starved to death; they were on gravel bars and untouched by scavengers. We can't be certain as to why they haven't been scavenged, but wolf, fox, and raven densities must have been low. We found a lone yearling living on a craggy bluff, on its own for two full weeks before it died. Again, no wolves. A three-year-old first-time mother birthed at the margin of her herd in late April. The baby could not stand, but the group stayed put, at least for one day. I move in to see if the neonate is still alive. She bleats. I put a finger out, and she tries to suckle, as if on a nipple. I feel like mom but can offer neither milk nor safety. We carry the newborn seventy-five yards to the waiting herd. The new mom stares helplessly, even after we leave the baby.

The next day the calf is dead, not eaten, not scavenged; this is natural selection. Despite humans' perception about nature's beauty, nature is also very unkind. Life becomes death—pets in homes, horses on farms, and innocent people in a bazaar: none are immune.

Nature and evolution are much the same, in that they do not always subscribe to the idyllic images or painless processes we might like to think of. An adult cow may lay breathing and motionless for days as wolves munch and ravens peck, with herd mates steadfast nearby. Among social ungulates, females tend toward higher encephalization quotients than males. Elephants are sentient. Females remember their dead. Do muskoxen?

It's late April. I'm standing on a knoll as mountains and valleys still dressed in their finest white linens sweep the horizon. Below, the earth drops a precipitous five hundred feet. On this windswept knob are my friends, with a hundred yards separating me from the herd. Three adorable dark puffballs are on the ground.

For a split second I think wombat—*thick-bodied, short-faced, powerful legs, short ears, small eyes, and tailless.* I work quickly, completing my photogrammetry assignment in two minutes. Downwind and away from this rooftop safety, less than a thousand yards off, are five animals. They're dark, but not muskoxen. And bears don't come in small herds.

These do. From my vantage on the knoll, I see no food for them. Then, I realize the meal is right here, right beside me—puffballs and their moms.

Pat Reynolds and her colleagues wrote a report in 2001 summarizing predation events by bears on muskoxen in northeastern Alaska. A few years later, in May 2007, Jeanice Walters posted a fascinating roadside video of a single collared grizzly taking out young calves (https://www.youtube.com/watch?v=-0ORjvSEYuo). That herd had at least two adult males, but all animals bolted, apparently without protecting the young.

Muskox herds normally contain adult males and females, a pattern consistent across Greenland and the high Canadian Arctic to Russia. There are always a few exceptions but, where not hunted, such as in parts of Siberia's Arctic, thirty-four of thirty-five groups had adult males.

Alaska differs in that most groups have fewer males. Muskoxen occur in areas where hunting is permitted, with some localities allowing quite liberal harvests; both sexes can be shot on the Seward Peninsula. Males are preferred by trophy hunters, but not by those hunting for the sake of subsistence.

The issue is this: groups without males contain fewer calves. No one knows why. One possibility is that those females are less fertile, either because there are fewer males around to provide sperm or because the females tend to be in poor condition and therefore conceive less, irrespective of male abundance. Neither idea has support.

We found pregnancy rates to be high and about the same at both study areas—Bering Land Bridge and Cape Krusentern—and the condition of the females did not vary between areas, despite the absence of males. The lack of babies must therefore be due to something else. Understanding this issue has bearing on the potential for population persistence: if the young do not reach adulthood and do not reproduce, populations will inevitably decline.

But back to bears, for the moment. Everyone knows they abound in Alaska. If you're Alaskan, you also know the polarization that accompanies their presence, especially state policy that enables liberal shooting and trapping of predators. Both bears and wolves can be good at killing the ungulates that numerous Alaskans subsist on. Yet, it's far from a zero-sum game whereby every ungulate killed by a carnivore deprives a hunter from putting wild meat on the table.

Many factors affect populations—body condition, winter and summer weather, insects, parasites, and disease. Nevertheless, the state of Alaska, just as many others in the Outside, including Montana, Idaho, Wyoming, and Colorado pay a lot of money to kill predators. Such economic decisions do not assure more ungulates for harvest, though that is often the intent. In fact, one of the past test questions to prospective Alaska state biologists in their oral exam was whether they'd have a moral compunction about killing predators. The question I'm particularly interested in is not about shooting carnivores—though that's relevant—but whether the hunting of adult male muskoxen is causally related to the lack of surviving youngsters.

Two Alaskan biologists, Josh Schmidt and Tony Gorn, suggest just this in their 2013 paper "Possible Secondary Population-Level Effects of Selective Harvest of Adult Male Muskoxen." Here's the supposition: males defend groups, and without males, carnivores and especially bears are more likely to kill young. Schmidt and Gorn present their hypothesis carefully. What's the evidence?

Across three decades, the number of muskoxen increased at about 15 percent per year on the Seward Peninsula. Between 2002 and 2012, for which the best information is available, abundance dropped locally at rates from 4 percent to 12 percent each year. Where males were heavily harvested by hunters, population declines were more prominent. If Schmidt and Gorn are correct, then implications for enhancing the longer-term survival of muskoxen will require a change in the harvest of males or, as others would prefer, in the killing of more bears. Some factors can be more immediately controlled, some cannot. The weather and *ivus* are among the latter; harvest and predators are among the former.

A closer look at assumptions is useful here. Chiefly, it would be helpful to know whether manipulation of adult sex ratios has deadly consequences for the survival of young. Social structure might matter, yet knowledge about the responses of individual male and female muskoxen to approaches or attacks by grizzly bears is virtually nonexistent. Males might have little or a lot to do with group defense. Basically, the tack—to guard or not—is about a trade off: either invest in immediate protection against a bear or avoid or delay defense, providing an assuredly higher probability of individual survival. The latter tack offers an option for future reproduction. Whether the bulls in groups are the actual fathers of the calves is an entirely different issue.

The Schmidt and Gorn dataset is based on overall changes in the sex ratio across time but not at the level of individual groups. So we don't know what the actual group composition was at the time the calves perished. We also don't know with much certainty whether bears are the primary source of mortality, but evidence from Pat

Reynolds and others seems solid, with wolves not playing a central role here.

Assuming grizzly bears are the primary culprits, the issue is layered with uncertainty because the behavior of male muskoxen remains a mystery. Perhaps groups attacked by bears have bulls, but they're wimps, employing the "wait another day" tactic. Maybe they are more likely to depart than females. If true, then behavioral differences between the sexes, and not overall sex ratios, have more to do with juvenile survival. And then there are the females: What do mothers and other females do when confronted by bruins?

Queries are boundless. Science works by testing assumptions and digging deeper. Although Schmidt and Gorn offer the best available data generated from aerial censuses, we know little at the ground level.

If males actively defend, does the intensity of their behavior change when calves are present? In other species, evidence shows that defense increases when individuals or their kin invest more in progeny production. Might muskoxen males lead attacks, followed by cows, or is it the converse? More centrally, how do groups without males respond to predators? Are females more likely to run when males are absent? If so, are attacks by bears then more likely to be triggered? Finding even a few answers will be a challenge.

There is a related way to approach the logic of discovery. If low juvenile survival truly reflects predation by grizzlies, then at sites with lower rates of population growth, the responses of adult females to bears should be greater than elsewhere. The logic here derives from the fact that mothers know their babies and have learned that predation is a cause for their death. I knew this to be true of moose mothers based on my past work, and so why would I expect moose mothers to be smarter than muskox mothers? I didn't. Consequently, I predicted that adult female muskoxen will vary in antibear behavior based on experience. If, for instance, predation exerts a stronger influence at say, Cape Krusenstern and Cape Thompson than on the Seward Peninsula, females will respond more strongly to bears in the north than in the south. Maybe bulls play a role, and maybe they don't.

To support or discard the idea that male presence is a deterrent would require contrasts of juvenile survival in groups with and without males when attacked. If females charge bears more than bulls do, or if bulls flee more than do cows, then maybe male presence is simply moot. Without seeing interactions and measuring these behaviors, many stories can be concocted.

Over the years, I've encountered grizzlies at the base of knolls, and on half a dozen occasions their footprints in snow lead directly uphill to knolls, some with muskoxen and other times empty. Only once did we observe a grizzly working its way toward muskoxen on a ridge. The sound of our snow machines caused the bear to abort any approach. It ran, terrified as it should have been, knowing that often it becomes prey.

Today, the five grizzlies below my muskoxen knoll are in climb mode, moving in my direction. I'm vulnerable on the football field–sized knob. Leave, and never learn the outcome? Or stay and put myself at risk?

That's a real problem when studying species in remote environments. One cannot easily witness interactions, let alone get into animal minds. Such dilemmas are sometimes resolved by using controlled field experiments. Two scientists did this long ago. In 1973, Niko Tinbergen and Konrad Lorenz shared the Nobel Prize in recognition of "discoveries concerning organization and elicitation of individual and social behaviour patterns." Each scientist used an experimental approach to develop an understanding of animal perception. I never imagined that I'd consider experiments in which I'd compare responses of animals to tape recordings of different sounds until I began working on predator-prey issues. While I had conducted sound or odor tests in the Arctic, I had not then worked with muskoxen and was uncertain how I would learn about muskoxen reactions to bears. The intense

Arctic wind would likely diffuse the perfumes—and, in turn, any detectable responses by muskoxen—to the bear poop.

My dilemma festered for nearly a year. I considered the possibility of a robotic bear to approach muskoxen and then to judge muskoxen responses. After all, the rovers called *Spirit* and *Opportunity* navigated remotely and crossed difficult terrain as part of the Mars Exploration Mission. They cost more than $800 million, but less expensive robots were available. An android sage grouse was used to approach hens on Wyoming mating grounds, and a sham badger on plastic tires assessed black-footed ferret temperament. In the Serengeti, customized stationary lions helped discover whether male mane color and size affected female mating choices.

No one, not even the National Science Foundation, was likely to fund me for Mr. Robo-bear. Logistics would be formidable—build the model, keep batteries charged, navigate snow and drifts and ice while powering up chiseled mountains. Its operation would have to be distant to disassociate a person from Mr. Robo-bear. Possible—yes. Inexpensive—hardly. Doable—hmm.

Nevertheless, getting inside the minds of muskoxen seemed worthy if I was serious about discovering the likely roles of males and females in herd defense. Long ago, others realized the value of animal models to hunt prey. The native Hausa of northern Nigeria acquired meat by mimicking ground hornbills. Disguised in feathered headdress they crept to unwary victims—antelopes—that had little fear of the ground-dwelling birds. When within range, the Hausa shot arrows.

Otto von Kotzebue described: how Native Americans hunted in California in 1824: "These [elk] generally graze on hills from whence they can see round them on all sides. . . . The Indians, however, . . . fasten a pair of the stag's antlers on their heads, and cover their bodies with his skin; then crawling on all-fours among the high grass, they imitate the movements of the creatures; the herd mistaking them for their fellows . . . are not aware of the treachery till the arrows of the foes have thinned their number."

I decide to pursue the use of a fake animal to determine if the manipulation of sex ratios affects relationships between muskoxen and bears. Moving concept to reality means convincing park supervisors the idea has merit and that I won't die. Approval comes with an understandable restriction: not until my results are published is either photo or other documentary material to reach a media lustful for unusual stories. The real work is about to begin.

Since I need a practical way to do experiments, I thought: Why not approach muskoxen groups dressed as a grizzly bear? Being my own bear would make it possible to approach every group of muskoxen in the same way, occasioning greater experimental control. Also, my donning of a bear disguise would make it possible to avoid problems encountered by the robo-badger and android grouse, neither of which could navigate difficult terrain. Ski poles in faux fur will be my front legs. I'll record data from a first person perspective and directly measure distances between the bear—which is to say, me—and muskoxen. I can gauge a range of behaviors, including charges by the muskoxen, measuring both their intensity and duration and recording whether the charges are made by females or males. Best, however, is that the bear model will enable systematic efforts to contrast group responses with males present and absent and, hence, to test group vulnerability to predation when sex ratios are biased.

Other issues need resolution. For instance, if I rely solely on the fake bear, I won't be able to conclude anything about muskoxen responses to bears—an odd thought because I'll be the bear. But therein lies the problem: the response of the muskoxen might not be to a bear but to something strange, maybe just a human or a novel object.

There will be no way to infer whether muskoxen view the approach as that of a bear or something else. A control is mandatory. I need an appropriate template for comparison. A model caribou makes the most sense.

Muskoxen see caribou in the wild, just as they see bears. One species is not dangerous, the other is. If my models work, then muskoxen should distinguish between them. I'll use a lighter-colored cape for the caribou and, as with the bear model, faux fur–covered ski poles as legs but will approach groups similarly. If species recognition occurs, muskoxen behavior should vary.

How I make my approach remains a concern. I can't just ride up to groups on a snowgo and present the two models. The noisy, smelly machines will have to be stashed downwind, probably a mile out. Fred and Wibke, the bio-tech, will be relegated to waiting patiently and will need to prep for serious cold to await my approach and return—maybe three hours of misery in wait time, total.

Issues of safety are also disquieting. What if I'm mistaken for a bear by a hunter or run into the real thing? At least mating season is not until June. For defense, I can carry pepper spray. Except that it doesn't work in cold and doesn't work accurately in wind. More likely is that I could be injured by an attacking herd, or worse.

Lethality is synonymous with muskoxen armament. The combination of a thick boss and sabered horns offers the delivery of a truck-like force with a skewered hook. At least one grizzly has died from a horn puncture. In Norway, where *moskus* (their local name) were introduced after World War II, an Italian tourist was injured in 1981 and a German in 2015. The deadliest attack was in July 1964. Ola Stølen, whose farm is in the Dovre Mountains, was curious and simply got too close. He was knocked down twice by a bull and died.

I wasn't sure what distance constituted foolishness but if I wanted to understand the mind of *moskus* I would surely find out. My friends reminded me neither they nor the government needed another Timothy Treadwell moment; Treadwell was the Californian who unwisely befriended Alaska's brown bears and was later killed by one. I consoled myself by remembering that the Dovre attack was not in winter. Maybe *moskus* behave differently in snow.

The first trial is near the pack ice on the coast. We kill our machines a mile distant. Four adult cows, two adult males, and five juveniles and subadults make for a nice assembly.

Most people cover a mile in twenty minutes, usually following a route on concrete or dirt. In winter they might snowshoe or ski it. My sojourn would differ. I'd not move directly from point *A* to *B*. Bears meander and sniff, which I would feign. And I'll posthole, the tiring process of walking and periodically sinking in deep snow, and overheat. This will be slow going. Under a long furry cape my daypack will create an aura of a grizzly hump. Strung across my neck will be a range finder and small camera; in my pockets I'll have a two-way radio, data book, and binoculars. Wind will be an issue, so I'll have my parka, goggles, a face mask, and fur-lined hat.

At 2:20 P.M. I bid good-bye to Wibke and Fred. In the event things go awry, they'll have to be ready. We weren't sure what "ready" meant. Fred cannot shoot at a mile. But I didn't need them zooming in and scaring animals if I was still experimenting. I'd call them on the radio if I was gored. My first step breaks through the iced outer layer of deep snow. Whoops. The trek is going to be long.

The herd is spread out in customary fashion, perhaps seventy-five yards across. Two youngsters are lying down, while the others stand or feed. At about three hundred fifty yards, an adult female notices my shuffling. She switches from graze to gaze. After eighty-seven seconds, she returns to her garden of frozen lichens.

My approach is indirect, that of a pseudo-forager with the tortuosity of a bruin. At two hundred seventy-five yards, the same cow decides to watch me full time. I walk parallel. Soon a second female grows vigilant. My goal is not stealth. But then again, there is no hiding in the Arctic. The charade continues.

At two hundred thirty yards, juveniles rise to standing. At one hundred seventy, the entire group stares. At a hundred fifty, the outermost animals join others in a group huddle. Males are looking. The wind whistles through the netting of the fake bear. My ears hurt as windchill registers about −15. I continue, convinced I can tough this out.

At about eighty-five yards, my ears lose feeling. I lie down, much as bears will. I re-don my rabbit-fur hat. It helps my numb ears. The animals are not puffing but starting to wheeze and swirl. At forty-five yards, I again lie down.

A male, defiant and confident, stares from the group's forefront. I'm reluctant to crawl closer. But I have to. The group whirls in uncertainty—spinning in a circular pattern with an ebb and flow of the young from the rim to the inside. Is this part of their effective defense?

I wonder about offense, otherwise known better for what it is—an attack. I think of Ola violently slammed near his Norwegian farm by a runaway locomotive. My breathing is deep. I'm afraid. I don't know why I'm doing this. I think of my daughter Sonja, but the thought is fleeting.

Fury and power is on its way, one male approaching fast; he's at thirty yards, twenty-five, twenty. No time to think. No place to run. *Really?* I'm not ready for impact. *Oh, fuck.*

It's now 3:25 p.m. Fred and Wibke duck the wind, watching neither me nor the herd. At this very moment, the herd and I are one and the same to them. Watching a lunatic sink in snow when walking even half a mile, rest, and then detour, hoping for stable snow, all to approach a group of wild animals at the edge of the world can't be very exciting. My trusted colleagues have no idea of the freight train with a death-dealing wish now barreling toward me.

In contrast, I am well aware of the bull's menacing approach but am helpless to stop the battering ram with hooked horns. That's what I think anyway. Then instinct and adrenaline set in.

I fling the bear's head skyward, jettisoning it twelve feet high. Its furry black back (cape) is airborne, though not quite with the finesse of a matador's swirl of muleta directed at an enraged Spanish bull. Not even close. I have neither the guts nor the skill for that. The senseless primate has tossed both his head and cape. Maybe now the amped Mr. *Moskus* will target something else. Wrong.

Instead, he halts. He has no target.

I imagine his confused mind: "Huh. What just happened—the bear, no: the human? WTF?" An ursine torment detonates, right behind insolent eyes. In the snow is a decapitated head, with its caped body to the left. In the midst is a fragile biped whose loud utterances sound much like the word "whoa," over and over.

I don't wait to see what happens next; my beeline back to Fred and Wibke has begun. Soon, there's a faint whir. They'd been watching after all. They're coming for me.

Over the next few years, I continue these visual presentations to *moskus*. Not all are on flats because I must find other herds. I run the experiments in mountains and in valleys. I do them on rocks and on permafrost, and with groups with juveniles and without. I do them to herds that have males and those that don't, and even to groups that are all male. In addition to behavioral responses, I record external conditions that might predispose a herd to remain, as well as its flight distance. I measure snow depth and the hardness of snow, noting how often *moskus* posthole for every hundred steps. If the surface is hard, are there more attacks or more flights? If the ground is torturous and animals must work hard to walk, such as when postholing is frequent, will they stand their ground? I also measure their visual field, noting how far until the horizon is obscured to them. Even with their relatively poor vision, if *moskus* are in a small valley, they won't see as far as on open tundra and perhaps will have little advance warning of an incoming enemy.

None of my information will make any sense, however, without my control. So, I need to do each experiment over again with the caribou model. At most I can get only two data points a day, and for many days there will not be a single one. Sometimes fog will come in and, because we won't see animals until we've gotten quite close, we'll inadvertently spook them with the snow machines.

On productive days there were two sets of experiments, and on a most fortunate day I scored four data points. I had approached one herd as a bear first, and the group became unruly; but in a few hours it settled down, and I later pretended to be a caribou. Sometimes I'd be a caribou first. Approaches were always a trudge, a mix of adrenaline and apprehension blended with sweat and numbing cold. But the most successful field season netted twenty-four data points. Over the years, the sample built to over one hundred.

A first cut at data was revealing. Female groups that lacked males were more than three times as likely to run. With males, they typically stood their ground, at least for a longer time. The findings presuppose that *moskus* considered me a real bear. Had the model been a real bear and not me as a bear, reactions would have likely been more severe, especially given my weight and size was only a third that of a real bear.

Most relevant is that when the results of "bear" were contrasted to those of "caribou," both types of groups—female only and female with male—remained more often when faced by the sham caribou. The *moskus* mind knew the difference. Though unsurprising, it was important to confirm the distinction and then build beyond to understand more fully how bears shaped the biology of *moskus*, at least in Alaska.

Other factors had also had an impact. Snow hardness affected how far animals ran but not whether they remained or fled. The presence of young had virtually no effect on behavior. Most surprising is that I was never charged by a female. This might have been the case because I did not cross some threshold, maybe twenty-five yards. But males charged. Of the four most serious, and of the seven other approaches when an individual left the group to come toward me, it was always a male. Given that females were at least eight times as abundant in groups, males were on the order of a hundred times more likely to attack.

That didn't mean females would not attack, and I did not discount that possibility. Back at the research center in Fairbanks, angry cows charged me several times, something I remembered when working with their wild cousins. In Norway, tourists are warned not to approach within two hundred meters. In Alaska, people approach much more closely. It goes the other direction too: muskoxen have been known to approach people. Berry pickers worry. Mushers worry.

The bottom line is that adult males play important roles when herds are approached by bears. Without males, groups are more erratic. Flight is a good way to invite an animal attack. We all learn this early in life: encounter an aggressive dog off leash, standing upright is not a good thing, but running is worse. In the wild, whether a bear is black or brown or white, running invites mauling. When males are present, my results tell me the defense of the herd will improve.

I admit to a bias. I want the Arctic's most majestic mammal—its longest lasting, largest lord of the land—to enjoy a rosy future. Am I blindly optimistic or pragmatically doleful?

Despite the slow accumulation of facts, uncertainness outpaces answers. My focus has been the muskoxen at the extreme western edge of their North America distribution, where the Chukchi Sea limits expansion. I asked questions about processes that affect population change and concentrated on juvenile development because nutrition governs individual growth. My enthusiasm for data collection shifted from radio-collaring and weighing animals to less intrusive methods because aerial assaults stress animals and relegate individuals to solitary confinement, often in snow holes. Instead, I intensified photo-imaging efforts to assess individual sizes and increased my collection of poop for pregnancy and stress diagnoses. Although my interests were spawned by the enormity of climate challenge—in terms of both temperature change itself and shifting ice regimes—they soon expanded to the vicissitudes of muskox life, given the inextricable link between the livelihoods of people and animals. I explored muskox social structure, because of its disruption, and the sex ratio skew caused by human harvest. My use of fake animals was to understand the mind of *moskus*, specifically the particulars of its vulnerability to predation by grizzly bears. Everywhere,

Arctic logistics, dovetailed with time, defined the extent of our ability to gain knowledge.

What did I gain from installing the radio collars, and the subsequent fiascos? Like most who employ GPS-like technology, I did find fundamental gains for science and at times recognized sacrifices that individual animals had unwillingly made. I also realized that understanding adult survival, and especially diagnosing causes of mortality, is rarely possible without a quick on-the-ground follow-up on individuals. Generating massive sample sizes to understand the nuance of how habitats are selected, or how and why animals move, is regularly facilitated by the technology. But stresses, disruptions, isolation, and death to individuals are the prices paid. My change of heart was fueled by watching terrorized animals flee and herds split apart by the heli-darting.

I asked my guides and other locals what they thought of radio collars. Some thought they were an important tool, but most did not. Many just shrugged. Freddy felt longevity for muskoxen in western Alaskan was dicey and remained unsure how much the collars helped. His belief was in part because of the severe damage of *ivus*, and especially because of rain-on-snow events, perhaps like the one killing twenty thousand muskoxen on Banks Island.

His perspective was fortified by yet one more dance with death. It was December 2014. Fred was thirty miles out when his snow machine just stopped. He couldn't fix it. Armed with his usual gear—tarp without sleeping bag, matches without wood, gas but no stove—he spent three days wrapped in canvas. There was wind and snow. Most notably, there was rain. Temperatures hovered between 22°F and 33°F, perfect for hypothermia. Near lifeless when discovered, Freddy was saved by the unseasonably warm winter temperatures.

For muskoxen, it's just the opposite. Warm and wet periods are challenging, while cold and less humid weather is conducive for survival. For me, after nine years, I began to understand rain-on-snow events and their nuanced but biologically important effects on mothers and unborn babies, food security, and accessible resources. Every human mother knows that gestating fetuses experience tremendous growth in the final trimester. We all appreciate the increasing appetite of soon-to-be mothers whose energy requirements magnify threefold or more.

Imagine that you're a pregnant mother and your only food is salad. The refrigerator ruptures, leaking water all over your salad, but assume the freezer kicks in and turns your dripping salad into an inedible chunk of ice. Now, transform yourself into an expectant muskoxen when a rain-on-snow event hits. You can't eat because your tundra salad is rock hard. Only body fat is available. What's the impact on your growing fetus? It's not good. Nutrient deprivation occurs. If the infant survives utero passage, it's born light, or worse—wimpy. What's next?

Fetal malnourishment has consequences, some lifelong or life shortening. Foremost, lighter babies have higher death rates. They tend to grow more slowly, can be compromised physically, experience lethargy, play less, and experience delayed sexual maturity. If small themselves at adulthood, they birth slighter babies in a cycle that repeats itself. When nutrition is compromised early, juveniles will be handicapped unless these initial effects are compensated for later in life. A mother's environment plays a vital role in her offspring's future fitness.

Over the years, I measured head-size growth in juveniles; I followed up by examining how seasonal weather events in different years affected the vitality of young and what, if any, extreme events occurred during pregnancy. Cindy Hartway, a talented biostatistician and WCS ecologist, helped. We accessed weather data recorded by satellite for my study areas and included measures of vegetation, as well as summer and winter temperatures, and of course episodes of rain-on-snow. We checked for possible relationships with the Arctic Oscillation, a driver of climate variability facilitated by counterclockwise winds across the Arctic, and with the Pacific Decadal Oscillation, a long-lived El Niño–like pattern of Pacific climate variability.

We found rain-on-snow stunted infant growth if occurring during pregnancy and for up to the first two years of post-birth life. We didn't know if the impacts were life long, but the effects of weather on muskoxen were clearly demonstrable through both *ivus*, which produced significant one-time mortality, and rain-on-snow events, which retarded individual growth. Coupled with David Gray's earlier work on muskoxen at Canada's Polar Bear Pass, where severe cold catabolized the energy resources of mothers and likely prevented conceptions, the wide variance of weather produced strong impacts at high- and mid-latitude Arctic sites.

We had other insights as we went along, some with alarming signals. Given the differences in population trajectory between the muskoxen at Bering Land Bridge and those from Cape Krusenstern north, I had expected animals at the former to be larger and more robust. Yet head sizes did not differ between study regions, nor was there variation in pregnancy rates. In other words, nutrition was not a significant factor. Something else must have been.

My attention turned to the altered dynamic of group structure caused by hunting—specifically, the skew in adult sex ratios and the resulting vulnerability of herds lacking males to bear predation. The changes in challenges to muskoxen persistence were numerous, some directly caused by humans, like harvest, and others with causes less direct, like climate. Emerging pathogens were apparently responsible for taking out nearly a quarter of our radio-collared females one summer, suggesting that disease played a significant role.

There was one more factor to consider, which was striking not only because we created it but also because of the consequence—namely, the solitary confinement that we had imposed on adult females when we failed to reunite them with herd mates after they were collared. Their subsequent cortisol levels were some three to five times higher than their undisturbed herd mates. We were able to determine the physiological stress but not the social or psychological effects.

The longevity of species at the world's highest latitudes will be affected mostly by forces that differ in both kind and scale from elsewhere. A case in point is the Arctic portion of Alaska. Larger in size than New Mexico and with less than twenty-five thousand people, its human density is some eighty times lower than America's intermountain West. Humans in the Arctic should not be an issue affecting the persistence of species.

But we do, and have, both indirectly and directly. Despite even lower human densities more than a century ago, muskoxen were lost to Alaska for one reason: people killed them. Without an ethos for species conservation matched by enforceable laws, persistence by a species is a gamble at best.

The Arctic is big and its trials different than anywhere else. Neither fencing nor the isolation of populations are issues likely to affect large mammals like caribou or muskoxen. In more crowded environs such as the Everglades or Yellowstone, where wide-roaming species face real constraints, genetic effects of population isolation are increasingly seen. Kinked tails and nondescended testicles in cougars and limb deformities in bison signal inbreeding, a problem that emanates from reduced dispersal. For ice-dependent species like seals and walrus and polar bears, other factors are pushing adaptive capacity; climate change is chief among them.

Northern landscapes, too, are being transformed. Lightning strikes are more frequent, and thousands of miles of tundra burn annually. Grizzly bears are more common on Arctic islands of central Canada. Vegetation change has prompted the northward movement of moose as habitat productivity increases. Caribou transplant pathogens and parasites from the south to the north.

The scale and time over which these documented metrics occurs indicate that conservation approaches used at lower and more crowded latitudes may be inappropriate across vast northern lands. The time required for vegetation recovery, for instance, cannot be measured in years but instead may take decades, or maybe centuries.

This means that, for conservation to work, planning efforts will have to incorporate broad swaths of land across long time periods, rather than the compartmentalized work practiced in the lower United States or other areas of the world; those are places where humans dominate and free-ranging large species are squeezed by competing land uses.

High-latitude challenges will relate to processes beyond the weather and the biological. Humans already insist on more roads, infrastructure, and digging in order to mine and develop additional energy sources; all have great potential to displace animals from crucial habitats. Harvest, if not closely managed, can indirectly alter muskox herd structure and tip the balance of survival to carnivores. But hunting is manageable. So are roads, mining, and exploration. Where not harassed, caribou, muskoxen, and other species habituate. They can flourish in areas with human infrastructure, whether aesthetically displeasing pipelines or wind turbines.

Critical pieces of the puzzle remain, including scales at which the currently changing temperature and precipitation regimes govern species persistence. If vast, widespread, and unfettered lands and seas endure, then immediate management may matter little in light of the plurality of climate impacts, even devoid of additional complications by humans.

I want the answer to a different sort of question: What is it that nations are willing to commit to sustain Arctic biodiversity? Once we answer that, then maybe we can put ideology into action.

PART II

Sentinels of Tibetan Plateau

Everyone knows Earth has two poles, one north, one south. Meteorologically, there's a third. With forty-five thousand glaciers, the Roof of the World harbors the greatest accumulation of ice beyond the two poles and generates its own weather. Unlike the magnetic poles, which have no permanent settlements and occupy a mostly flat landscape, the Tibetan Plateau casts direct shadows on the third of the world's people living below it, who are fully dependent on its function as Asia's water tower. It's in these distant uplands that rapidly warming temperatures test human livelihoods, as well as those of a suite of cold-adapted specialists.

Extreme elevation limits all life. Yet here, in the tallest of the world's mountains and plateaus, more than two dozen endemic mammals persist. There are giant yaks, squat foxes with luxurious fleece, and wild goats with ornate corkscrew horns. Fat-bodied antelopes with sabered weapons mounted on their heads, alluring cats with profusely spotted fur who are a tenth the size of snow leopards, and desert-adapted brown bears all rely on the arid grasslands atop the permafrost in this place. Here, among the world's most punishing conditions, climate magnifies the environmental factors affecting the Asian wildlife and peoples in ways similar to the perils now being experienced by inhabitants of the polar north.

> The most remarkable [feature] of this landscape [was] its enormous elevation... further northwards, the mountains were rugged, often rising into peaks which from the angles I took cannot be below 24,000 feet and are probably higher.
>
> JOSEPH HOOKER, *Himalayan Journals*, 1854

7

Below the Margins of Glaciers

I head west from Beijing. My end point is fuzzy. Beyond Lhasa, the journey will truly begin somewhere in the great abyss to the north. For the fourth time I'm stopped by the authorities.

"Visa? Why you go to Tibet? You have books? Check luggage."

Despite sponsorship and subsequent approval from the Tibetan Forestry Bureau, my neatly braided long hair is not a likely asset. Nor are my three hefty duffels the normal luggage of a conventional tourist. My claim to be a scientist raises more doubt. Unflinchingly, I produce documents. The inquisitive sentinels search. The contraband they uncover is nothing more than my Swiss Army knife.

The secrets of Tibet—land, culture, political position—have intrigued Westerners for centuries. Chinese sensitivities

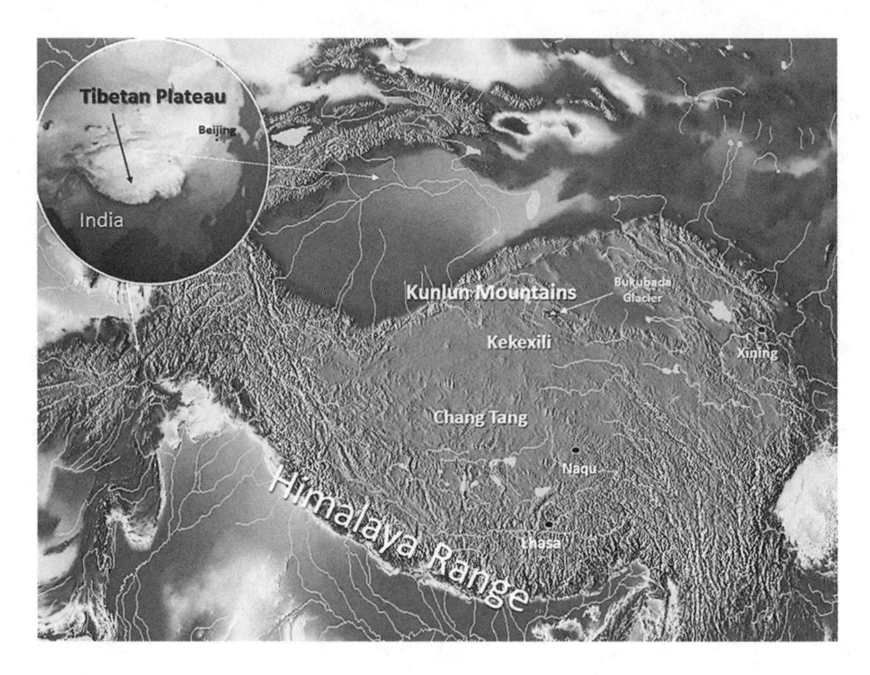

flare regularly. In 2010, it was no different. This time those sensitivities reignited in Lhasa because of increasingly stern policies to disaggregate protesters following half a dozen monks and other Buddhists setting themselves on fire.

The twenty-first century is spun backward in time, as if the Great Game—when China, India, Russia, and England wrestled for control of the inaccessible highlands, among other Central Asian areas—were as yet ongoing. Maybe it never ended. Unrest and wars about sovereignty and ideology seem never to dampen in or beyond Afghanistan. Tribalism and patriotism spawn suspicions about neighbors. As the planet's population spirals upward, the lethality of human weaponry grows. So does the grab for limited resources.

In 1774, French missionary George Bogle traversed the great plateau just north of the Himalayas, noting similarities to the polar north. "The coldness of the climate renders fuel a very essential article as no wood is to be had." Just as Inuits used local goods—

whale blubber and seal fat to generate warmth—Bogle discovered similar ingenuity: "Tibetans are obliged to use cow dung.... It makes a cheerful and ardent fire."

"No yak, no Tibetan people," is what the 10th Panchen Lama said, realizing human life among the inhospitable highlands could not exist without domestic yaks. Other travelers reveled in the mysterious plateau and its life forms.

A century after Bogle, Robert Shaw wrote: "Those who penetrate into the wilder parts of Ladak bring down reports of the wonderful animals. . . . Wild sheep as large as ponies, wild cattle with bushy tails like horses and long hair on their flanks reaching nearly to the ground, besides antelopes and gazelles." In 1895, Hamilton Bower, a British military officer, pondered "how these animals live through the arctic cold of the long winter, when the country is covered with snow, hiding the little withered grass." He considered it "a mystery [that] can only be accounted for by the amount of fat they lay up when the grass is at its best." Ten years later, Oscar Crosby offered the following account: "We travelled for eight weeks, never at altitudes less than 15,500 feet, often rising to 18,500 feet. The country is quite barren and uninhabited, and the cold is extreme."

But the Russian explorer Nikolai Przhevalsky had delivered a contrasting opinion in 1876: "Notwithstanding their sterility and the unfavourable conditions of climate, the deserts of Northern Tibet abound with animal life. Had we not seen with our own eyes it would have been impossible to believe that in these regions, left so destitute by nature, that such immense herds of wild animals should be able to exist, and find sufficient nourishment." He added, "They have no fear of encountering their worst enemy, man; and far removed from his bloodthirsty pursuit, they live in peace and freedom."

Among the species noted for docility were chirus, wild yaks, gazelles, and two wild equids—one half ass and half horse, known in Latin as *Equus kiang*, and its cousin *Equus hemionus*. In English, these are called kiangs and khulans; the former is an extremist, gorgeous in its mix of dun, tan, and beige, living mostly above fifteen thousand feet and solely on the plateau. Khulan, its low-desert relative, is an equally stunning wild ass of the Gobi, though it extends far beyond through India's ferociously arid realms to Turkmenistan and Iran.

Explorers used species behavior to judge relationships with people and remoteness. In 1898, British captain Montagu Wellby provided a description: "Here we saw immense herds of antelope, all females and young ones. They were so timid that it was impossible to get nearer than 600 or 700 yards to them.... We concluded that as thousands and thousands of wild antelope had chosen this pastureland to live in, surely it must be a befitting place also for nomads and their flocks." Three years later, Henry Deasy wrote that the lyre-horned chirus "were exceedingly tame and.... It was very evident that they were quite unfamiliar with human beings."

Sven Hedin similarly reported in his 1905 account *North and East Tibet* (the third volume of his *Scientific Results of a Journey in Central Asia*)that wild yaks and khulans in areas without humans "were so little shy and so numerous that, I concluded, we were still a long way from the first of the nomads." The intrepid Swede also saw "herds of wild yaks in several directions, but all of them just at the lower margins of the glaciers."

Today, the high mountain topography is unchanged and its wildlife remains unfamiliar to most outside this unfrequented domain. The Tibetan Plateau (also called the Qinghai-Tibet Plateau) varies from about ten thousand to more than nineteen thousand feet with summits rising to nearly twenty-four thousand feet. Roughly 50 percent is above fourteen thousand eight hundred feet. It's cold, arid, and bombarded by stratospheric-force winds. Temperatures still drop below −40°F. And, like the planet's high latitudes elsewhere, the Third Pole has opened its doors to warming temperatures, glacial recession, and challenges from humans. There is poaching, food competition between wild species and growing domestic flocks, and mining for energy or minerals.

The Tibetan Plateau spills out from China into Burma and Bhutan in the southeast, and Nepal and India in the southwest. Its western end nudges the mighty Karakorum and Hindu Kush Ranges into Pakistan and Afghanistan. If situated in America, its land coverage would reach from about the Mississippi River to the Pacific Ocean and from Mexico to Canada, yet the vast area in Asia is more diverse. When the legendary biologist George Schaller first began work there in the 1980s, the word "conservation" was unfamiliar to Chinese authorities. "Their eyes glazed over when talk turned to protected areas." No longer is this the case.

The plateau's biggest national park, the Chang Tang, is larger than Colorado. The adjacent reserves of Kekexili, Arjinshan, the twin Kunluns, and Sanjiangyuan comprise a protected network of unimaginable dimensions, equivalent in size to all of Kenya, or Montana and Nebraska combined. Human population densities here mirror the Arctic; there is less than one person per square mile. Chunks the sizes of West Virginia are unpeopled. More than forty Yellowstone-sized parcels could fit in Chang Tang alone. If one were looking for a lonely landscape outside the earth's polar zones, the Qinghai-Tibet Plateau would be it. Like many monikers, confusion may emanate where the past meets the present. Qinghai is a province with parts that connect to the plateau, and it has been and continues to be the home of different cultures, as outlined below. Across the plateau, human densities increase in locally attractive or incentivized areas.

Six Chinese provinces constitute the plateau: Gansu, Guizhou, Qinghai, Yunnan, the Tibetan Autonomous Region, and Sichuan. Collectively these harbor varied ethnicities, including Tibetan, Uyghur, Mongolian, and the Chinese Han. The people, topography, and physical challenges span a broad range of beliefs and politics. Economics and lifestyles affect the current diversity of wildlife and challenge its future.

Across the plateau, precipitation follows a gradient, with a minimum of less than two inches annually at the northwestern arid edge.

In the southeast, summer monsoons deposit more than sixteen inches. Unlike the Arctic with its winter snowfall, Tibetan glaciers are rejuvenated by summer's monsoonal rains, which drop snow at high elevations. Parts of the northern plateau also capture Siberia's winter moisture. The twin forces of longitudinal and altitudinal gradients shape plant communities, which are further mediated by permafrost and soil.

Below glaciers are meadows thick in turf-forming sedges. These are more widespread along the plateau's wetter eastern side and, throughout, are replaced by steppe grasses or deserts as elevation and precipitation moderates. The plateau is thin in plant coverage, with ground cover typically less than 20–30 percent. Seldom is habitat productivity even half that of the rich tropical or temperate grasslands of Brazil, Tanzania, or North America. The wildlife, however, is surprisingly diverse.

Still in existence are the species present during the expeditions of Bogle, Przhevalsky, and Hedin, who found some twenty-five or so large mammals as impressive as those in the well-studied Serengeti or the lesser-known Beringia. Eleven are deer, most associated with forests penetrating the plateau's southeast, including red (the same species as elk from North America), white-lipped, and musk. Along the southern boundaries and forested cliffs are goat-antelope derivatives, the goral and serow. From north to south, snow leopards are sustained by prey that changes from ibex to blue sheep. In alpine regions and desert steppes are Tibetan and Przewalski gazelles; at lower elevations live black-tailed gazelles. Between sand dunes and mountain peaks, argalis and chirus, as well as wild yaks and kiangs, can be found. Dholes (a canid) have hunted blue sheep at fifteen thousand feet. Lynx, wolves, and brown bears fill out the broader suite of large carnivores. For most, the species distribution has been identified, but not much is known of their ecological relationships. Pikas are the exception here. American biologist Andrew Smith and Chinese colleagues have dedicated decades of research and their efforts have revealed a keystone role for this attractive rabbit relative. Pika burrows aerate soil, allow greater percolation of snow and rain, and support diverse animal communities, including wild foxes, cats, songbirds, and avian predators.

For most field-workers, entry to the plateau is difficult. Restrictions along the western and northern plateau due to military sensitivities, remoteness, and a lack of roads all prevent access. Summer is sodden due to permafrost melting, monsoons, and blockage by vast salt lakes and runoff. Despite cold and wind, winter is the most accessible period. The ground and lakes are frozen hard. One can drive across the high alpine deserts.

During World War II, Franklin Roosevelt considered the feasibility of building a road across the southern reach of the Tibetan Plateau. Its purpose was to connect two American allies, India and China, allowing goods to flow to China that would slow Japanese momentum in Burma. The reality of staggering difficulties quashed the dream. Instead, the United States flew thousands of supply missions over "the Hump," an area where the Himalayas grade into the great plateau. Today, paved roads and a railway cross southern Tibet.

Unlike many early ventures, my business would be neither political demonstration nor mapping secret routes. I was intent on assessing wildlife above the deserts and exploring the possibility of a later project.

My assessment follows efforts begun decades earlier by George Schaller, who noted that the magnificent chiru may have numbered a million or more. By the early 1990s, some three hundred thousand had been poached, primarily for trade in the garment industry, in which pelts were smuggled to Kashmir. There, the fine wool from the pelts was woven into expensive shawls called *shahtoosh* and marketed in fine boutiques from India to England, and from France to New York. The entire population of chirus may have dropped to fifty thousand.

Likewise, wild yaks were slaughtered by road crews, by miners working illegal claims, and by military and organized bands bolstered by corrupt officials. No one ever knew the historical numbers of wild yaks. Like bison, the species collapsed; wild yaks survived only in the most remote areas. Argalis—the world's largest wild sheep with runaway horns that curl to its eyes—also disappeared from much of its Tibetan range. Across the region's lofty heights and down to lower elevations, the list of threatened and endangered mammals is startling—chiru, Bactrian camel, snow leopard, khulan and kiang, *takhi* (a.k.a. Przewalski's horse), wild yak, and saigas. Recent history has not been kind.

Although protection has come with the unification of massive dedicated areas, it carries costs and uncertainties. More people have relocated to areas bordering park boundaries at higher elevations, while others ignore the demarcations altogether. Fencing, mining, and road building continue. Only some of these projects are sanctioned. Bears are attracted to settlements for easier pickings. Together with snow leopards and wolves, they encounter abundant domestic flocks, which are not wily like their native counterparts—wild sheep, wild goats, and wild yaks. Conflicts with human and retaliatory killings of the wild carnivores grow, as increasingly tamed herds are encountered by the carnivores.

Among the most pressing issues is the changing nature of weather and the livelihoods of people living alongside wildlife. Tibet's dual traits of aridity and high elevation render its fragile systems susceptible to even subtle changes in temperature and rainfall. Across the last forty years, warming has averaged 0.5°F per decade, twice the global mean. This increase directly affects permafrost and snow melting. Of nearly seven hundred glaciers monitored by Chinese scientists, 95 percent are releasing more water from ice than accumulating it as snow, especially along the plateau's southern and eastern edges. Here, increasing soot and brown clouds generated by people on the subcontinent increase the rate of temperature changes at high elevation. As groundwater tables drop at the source areas of rivers like the Yangtze and Yellow, condition of the tundra worsens.

Inevitably, the pollutants enhance warming and lake levels lower, wetlands dry, and grasslands shrink.

Other disruptions to key ecological processes fuel additional conflict between people and wildlife. Some of the discord is less direct than bears breaking into the larders of semi-nomads, or snow leopards killing sheep. Rather, it is expressed through reduced grassland productivity, which subsequently amplifies competition for good pasture. Ultimately, meat and milk yield for domestic yaks is compromised for the very people reliant on the high altitude grasses. Increasing pressures are especially evident across the entire plateau, where more than 50 percent of the grazing lands have been degraded by high livestock pressures. In the western Chang Tang, human and livestock populations have tripled in three decades. Fencing accompanies the resettlement of people, and with more obstruction, the gate essentially closes to migratory pathways that allow wildlife to survive.

Like Beringia, the Tibetan Plateau is chilly. What Beringia once was during the late Pleistocene—with its camels, horses, and saigas adapted to aridity—can never be repeated. Of all areas on Earth, the plateau contains the most similar habitat for such faunal reincarnates. Its deserts and alpine steppes support species or their relatives that once roamed Beringia.

Instead of bison, the plateau has wild yaks. Instead of caribou, there are migratory chirus, and argalis rather than Dall sheep. Kiangs substitute for Yukon's extinct wild horses. Life in the aftermath of glaciers means subsistence in lands shaped by powerful outwash streams, erosion of soil and vegetation, and thick deposits of a dried silt-like substance called loess that has been fanned across the land by potent katabatic winds. This glacial dust permeates the fur of wild species, clogs modern vehicles, and pelts both herders and scientists.

Like animals, species of wild plants have occurred in both areas, some still do, and others have been swapped. Kobresia—a genus with numerous species of sedge was once in Beringia; it still occurs on the plateau and elsewhere in Asia. It's currently a favored plant in the diet of wild yaks and is also consumed by saigas down in the Gobi Desert. Kobresia was probably eaten by Beringia's bison and likely by saigas when they lived in the Beringian parts of Alaska and Canada.

The largest grazers outside the Asian tropics are wild yaks, and their traits parallel its New World counterpart, the bison; each is adapted to the cold and steppes. In 1790, Thomas Bewick, referring to the wild yak, wrote: "The Sarluc, or Grunting-Cow of Tartary, from its resemblance to the Bison, may be considered as belonging to the same species." This resemblance is of course a primary reason why Canadian agricultural authorities purposefully hybridized bison with yaks—to craft a better beast of burden. In the wild, bison and yaks have analogous social organizations that include the seasonal segregation of adult males from mixed groups of females. And both have correspondingly sad histories of slaughter.

There's also a big difference, however. Wild yaks have had a far more limited range than bison. Just as muskoxen appear to be a true cold-adapted species with both modern and past analogues associated with glacial retreat, wild yaks, too, have had a limited distribution. Theirs once extended from the Tibetan Plateau into India's and Nepal's upper reaches, across the Altai Mountains ringing western Mongolia, and up into the Baikal Region of Russia. These cold spots are typically conducive for yaks because they lack deep snow cover.

Unlike wild yaks, the distribution of North American bison was expansive, from the hot grasslands of Mexico's Chihuahua Desert and humid Atlantic forests, to arid spaces from eastern Oregon to west Texas. They occurred up into Canada's northern prairies and its boreal forest, even managing to persist in Alaska into the eighteenth century. Whereas bison had a penchant for a wide range of ecological conditions, wild yaks never did. Bison tolerate heat, at least up to 110°F without shade. None of this plasticity is incumbent to yaks. Yet which have been evolutionarily more successful, wild yaks or bison?

My efforts will involve more than just counting yaks and chirus, as important as such counts are. I want to understand the importance of recently deglaciated zones and warming to wild yaks.

It's late November when I travel north from Lhasa heading for Chang Tang, with my Swiss Army knife back in my possession. With me is Aili Kang, a field savvy, no-nonsense biologist who leads WCS's Tibetan office. She grew up in Shanghai, fled the megalopolis, and studied saiga behavior in a facility where they are captive in Gansu for her doctorate. She later joined George Schaller's 2006 expedition, along with ten Tibetans and two Han Chinese, to retrace the sixteen hundred–kilometer route of Captain Wellby, who crossed the plateau in 1896.

George, Aili, and the others would see no people and carry all amenities, petrol included. In addition to the usual challenges, they navigated a lattice through some of the plateau's thousands of saline lakes. The largest is at an altitude 15,480 feet and is seventy-five hundred square miles in size. Many smaller ones exceed five miles across. Countless lakes have no outlets—rivers or streams. Others, due to salt accumulation coupled with differences in water depth and ice thickness, are tricky to navigate. The group's journey began with four vehicles—two trucks and two land cruisers. Three completed it. The other found a watery grave after the ice gave out.

Beyond Lhasa, Aili and I say goodbye to free-flowing streams. We pass a few disfigured agricultural fields, and tractors dangerously overloaded with grass cane inching their way up steep mountain passes. Rounding a switchback, a truck sits limp, overheated. A dozen or so pilgrims climb down, smiling, and build a fire for a tea break. Farther up the road are more pilgrims—shuffling forth, some spinning prayer wheels inscribed with the Buddhist mantra *Om Mani Padme Hum* in Sanskrit. Others display colorful prayer flags symbolizing sky, earth, wind, and water. Rooted firmly in the cycle of birth, life, and inevitably death, the wayfarers go prostrate on the ground reaching with outstretched arms. This they repeat, over and over.

Lhasa is at 11,975 feet. We're now at fifteen thousand. We stop at the security station in Nagqu, where we register and pick up a Tibetan policeman named Baima. He seems nice enough, speaks no English, and will be my requisite guard 24-7. I wonder if my shadow is to protect me or report my activities. Baima is a lifetime smoker, and I wonder how he deals with hypoxia. Like other Tibetans, the secret is that his blood vessels are wide. This adaptation enables more generous oxygen flow to body organs and reduces the typical altitudinal sickness of many climbers and Han.

Back home in Missoula, Montana, where I then lived, I worried that I'd be plagued here with pulmonary edema, because the local mountains are picturesque but unimpressive in height, thus leaving me unacclimated. To prep for my journey, I pounded the hills, running in hiking boots, carrying two gallons of water. I was truly a pathetic sight, but after a bit, I easily navigated the high point at fifty-two hundred feet and raised my heart rate, although this still didn't take into account the reduced oxygen intake I'd experience by training at more lofty realms.

If we find wild yaks, I'll have to adjust. Aili has seen them at nearly at eighteen thousand feet, and chirus at only slightly lower elevation.

I wasn't sure what to expect generally, but historical accounts extending back to the eighteenth century and Schaller's books had given me plenty to think about, geographically, biologically, and culturally. Understandably, customs differ from culture to culture. Upon meeting my first nomads, I averted my eyes. They didn't and, in fact, outright gawked. Aili speaks Mandarin and English. The nomads speak only Tibetan, as does Baima. Our driver speaks Tibetan and Mandarin. So if I were inclined to ask a question, the process would go something like this: I'd speak English to Aili. She'd move from English to Mandarin, and then our driver from Mandarin to Tibetan. The nomad would answer in Tibetan, and the procedure reversed to Mandarin, and then to English. Baima would intervene as needed. This is standard practice—thirty minutes for a simple Q&A that might contain two basic questions. Patience is required in the Arctic, so why would the Third Pole be any different?

Something unexpected occurs in one of these encounters. In a simple, ritualistic, nonsmiling gesture, a nomad sticks out her tongue. It's directed at me, the obvious foreigner. Aili has not seen this before. Others have. Sven Hedin photographed tongue greetings more than a century ago. In November 1943, Robert Crozier and crew bailed out of their World War II plane after it was blown off course above Tibet somewhere between Kunming in China and Jorhat, India. They survived the parachuted downfall. Ciphers of friendship met them: outthrust tongues.

Some customs are more transcontinental. One steeped in spirituality, symbolism, and prowess, and maybe even machismo, is display. Wild male yak skulls—massive, horns outstretched, with eye sockets pointing to the heavens above the roof of the world—adorn mud-and-brick dwellings in parts of Tibet. Lisu hunters in Myanmar hang the heads of their kills at rural homes. In Patagonia, skulls of introduced wild pigs and antlered red deer greet visitors to people's homes. Sharpshooters in Wyoming and Colorado titivate ranches with their biggest of trophy elk. Deeper in time, mammoth hunters may have displayed their proficiency by hanging ivory tusks. Today's urban prowess is a different sort of display, one reflective of monetary worth—a Maserati, Jaguar, or an even bigger guzzler called a Humvee.

Other customs are universal and characterize both sexes—smiling. Aili does just this as she cracks open the local chang, otherwise known as beer. She does it with contraband, my Swiss Army knife.

> [The wild yak]... is able to defy the loftiest altitudes, the bitterest cold, the most violent snow-storms.... When it gets too warm for him he takes a bath in the nearest stream and climbs up the mountains to the cool expanses of the snow-fields and the curving hollows of the glaciers.
>
> SVEN HEDIN, *Through Asia*, 1899

8 The Ethereal Yak

Two yak caravans prepare to cross a high plateau. Amid swirling loess, the hazy sun disappears behind a rampart at eighteen thousand feet. Along the lower slopes are slabs of rock, one of which carries an etching of an animal with a bushy tail and a dangling skirt. The *drong* is well known here. Outside of Tibet, it goes by the name wild yak.

Untamed beasts come in unimaginable forms—fish with legs, bats that walk, moths the size of hummingbirds. Those beasts for which we are prone to add the phrase "of burden," though, are a less diverse lot. These domestics come from biomes tropical and warm, desert to grassland, and taiga to tundra. Their differences are as strong as they are obvious: they are providers of labor—hence the "burden"—and suppliers of meat and hides, footwear and fiber, and milk and curd. They cross time, cultures, and economies.

What reindeer, horses, and camels have in common beyond serving our immediate needs is deep ancestry. All were social animals, just like other domesticates, including sheep, goats, and pigs. Once wild and now tamed, they share not only the fact of an altered genetic makeup but also the role of feeding or transporting us.

Ecologically—and this is the key point—free-ranging biological ancestral forms are not especially popular today. Wild reindeer have been vanquished where domestic ones occur, both in Scandinavia and Russia. Wild camels are not tolerated by herders. The agriculture industry in the American West deplores wild bison; Canada's and Mexico's agricultural sectors feel similarly. Sheep ranchers loath bighorns. African farmers dislike zebras. Asian pastoralists want neither kiangs nor khulans. The antipathy is utilitarian; wild ancestors are suspected of bringing disease, which is typically an inarguable position. They must compete for food and steal females.

Such claims underscore a societal dilemma. Do we wish to protect native predecessors? Do the tamed animals that we've created now jeopardize the persistence of remnant ancestral forms?

Domestic yaks are an interesting case. More than fourteen million span treeless steppes from southern Russia and Kazakhstan, to Kyrgyzstan and high-elevation Bhutan, India, and Nepal. Nowhere is their abundance greater than the heights of western China where for thousands of years they've defined human culture. There are yak-headed gods and statues and yak blankets, tents, shoes, and other coverings. In the 1950s, the words "yakety yak" were made famous by the rock group the Coasters. A half century later, Yakety Yak became the moniker of a Colorado start-up business in digital communication. Yak burgers are now found in Bhutan, and Macyaks in Nepal. Yaktrax are outdoor footwear for bipeds on packed ice. Far from the domestic herds, the horns and skulls of wild yaks gather dust, grow white, and hang from mortar walls—tributes to society and long life. Yet few people actually think about wild yaks.

Ask someone in the Americas about wild yaks, and they will respond with surprise: "There are wild yaks?" Or: "I didn't know wild yaks exist." More typically, their eyes will glaze over. On the rare occasion that wild yaks are in the media, reports are usually accompanied by colorful portraits of domestic yaks.

Today's confusion about wild versus domestic is reminiscent of an incident that took place a century ago, when Sven Hedin had his own mix-up about the two: "The next day going down a valley, between granite mountains... a herd of yak grazed on a slope. Islam took a shot at the animals at long range; but they never budged. Instead of that, an old woman came running forward, shouting.... We learned from her that the yaks were tame." Earlier, in 1898, Captain Wellby praised the wild bounty: "On one green hill we could see hundreds upon hundreds of yak grazing; there was, I believe, more yak visible than the hill.... The fat of the yak was so precious to us that we used to boil down every ounce of it, and put it in our old cocoa tins... and eat it as if it were Everton toffy."

A crescent moon punctuates the dark sky. By the time we load the Land Cruiser and scrape our way along a poor excuse for a road, it's almost light. We cross stony valleys with deep clefts and rivers plated with ice. The moon fades as the sky shifts from crimson to sapphire. An hour later, a slanting sun casts long shadows on yak patties, all stacked neatly. The few herders who remain are readying themselves for winter, collecting yak dung for their heating and cooking. Meadows are thick with ice. A lammergeyer lifts off. We swing east, and the sun is now directly in our eyes.

At 30° from the equator—about the same latitude as northern Florida—the outside air is chilly. Geographical similarities are few. While nomads, like farm workers, appreciate the sun's warmth on cold mornings, two thousand miles north of Earth's midsection there are no workers, no crops, and little vegetation. The lakes are solid ice except for a few that are very brackish. Florida was once equally untamed, with wild antelopes, horses, and even mammoths. All of those species were lost, however, because thousands of years earlier the climate had warmed.

Here, on the high elevation of the Chang Tang, groups of Tibetan gazelles drink from shore, unafraid. Kiangs dot the arid valleys, sometimes in groups of half a dozen and sometimes a couple of hundred. Wellby spoke of their beauty: "[The kiang] is of a light brown colour, with white throat, belly, and legs; rather heavy in neck and shoulders, he is nevertheless, a graceful mover and there are few prettier sights than a herd of them scampering over some wide plain." If not for the nearby glaciers soaring above twenty thousand feet, one might be lulled into believing oneself to be in a tranquil African desert with slender gazelles and an evocative equid striped black and white.

A few hours pass. We talk with a herder named Lobsang, the sole person we've seen today, besides one other. I'm keen to learn about differences between the wild and domestic yaks. The latter are piebald and white, reddish, silver, or gray. At times, these variation occur in muzzle, wither, and chest.

Elaborating on yak husbandry, Lobsang says: "I was out with the yaks [*dri* or *dzo*—female domestic yak in Tibetan]. It was summer. A few days were cloudy, and sometimes it rained, but it snowed more often. There were many females, and I needed to keep watch because of wolves." I want to interrupt and ask how frequently he sees wolves, but I don't, knowing an unwieldy translation will follow.

Lobsang's ancestors were nomads, once living in yak-hair tents, and then became semi-nomads. He is more sedentary and moves his domestic yaks to new pastures less often these days. He's also a father. His wife milks the yaks, makes yogurt and cheese, and churns butter by hand. She offers me yak-butter tea. Globules of fat and hair coat the cup. I can't pour it out and thank her for my second cup. Lobsang continues his account.

"A *drong* [wild male] was on the nearby hill. I chased it away before, but it kept coming back. This time he came into the herd and chased my bulls. Some [of my] females were in heat. The year before *drong* came and took *dri*. I run out. The government does not let me have a gun. I try to scare it. He knocks me down. I was bleeding and my leg hurt."

Lobsang has a permanent limp. Farther afield, two herders were actually killed by wild yak.

Outside, Lobsang's brother and uncle saddle their yaks in flamboyant yellow and azure. The skies are jet blue and contrails never appear: jets do not fly here. Despite scarce resources, the herders manage to survive, thanks to their fathers' teachings. It behooves me to tread lightly, as I pursue my research, in this land the herders view as sacred.

The next day, we follow a rockier path to search for wild yaks. The terrain grows mountainous, and the scoured slopes reflect the remnants of long past ice floes. Boulder fields pockmark meadows. At sixteen thousand feet, we walk in snow.

The tracks are unmistakable: the deep imprints of brown bear. The nearest trees are three hundred miles away. Should I worry? Brown bears have killed more people here than have yaks. A Tibetan fox, more curious than afraid, stares. A dozen speckled snowcocks search the duff for food. Chirus have passed this way, leaving pellets as a telltale sign. In the distance are wild yaks. From what we can tell, there are three adult males and maybe fifteen females, with assorted juveniles. At three miles off, we hesitate to get closer.

They're skittish, maybe a holdover from days of past persecution. I wonder how long such effects linger. Testimonies to previous abuses may be found in behavior. On the domestic front, for instance, dogs let us know when we've been unkind by cowering or tucking their tails between their legs. In the wild, yak populations of those that were once persecuted remain wary for generations afterward. When encounters between wild animals and people are benign, animals will tolerate humans. Tibetan monasteries are well-known safe havens, for humans as well as animals: gazelles, blue sheep, and deer become unwary. I wonder if the current distribution of yaks reflects past hunting or simply the sculpting of land by snow and glacier. These are questions I'll explore in a subsequent trip.

Days later, we encounter another herder. It's cold inside his mortar home, maybe 40°F. A woman with high cheekbones and deep brown skin sits in thick fleece, wearing a trilby of colorful orange. She adds sheep and yak dung to the leaky stove. Wisps of particulates add to the decades of smoke on stained walls. Bashfully, she retreats behind her husband, who has gleaming white teeth and deep black hair. A necklace of turquoise hangs low on his chest, and a large conch studs his earlobe.

Baima explains we're counting wild yaks and chirus. The herder's children gaze at me, having never seen flesh my color. I motion, letting them know it's OK to high-five me. They laugh and slap, then touch my arms, laugh, and run away. Their father doesn't know anything about America, oceans, or what it takes to reach my confused part of the planet. He knows his backyard, the salt lakes, the snow regimes, and wildlife. He knows warming temperatures are changing grassland productivity on the terraces where his livelihood grazes—the yaks. I decline the offer for more yak-butter tea, now realizing the easiest path to declining is being certain there's no room in my cup for a refill.

Outside, stone walls confine animals at night. The walls do little, however, to protect them against bears and snow leopards. The herder's mastiffs help, but last year, a brown bear trashed their home when they were away. A few months ago, a snow leopard killed a calf; elsewhere, a two-year-old yak fell victim to a snow leopard. The family now lives farther out on the steppes, so the dappled cats prey on blue sheep in the mountains instead.

In order to understand local perspectives, Aili asks more questions, especially about small-scale fencing to help reduce conflicts between carnivores and people. Tibetan herders understand these kinds of conflicts, having survived the spectacular high barrens for thousands of years. Aili, of Han descent herself, points out that incoming Chinese have little familiarity with wildlife and hardly any desire to be herders. Conflicts between humans and wildlife may increase until there are fewer carnivores, fewer kiangs, and fewer wild yaks.

On the eastern side of the Qinghai-Tibet Plateau, wild yaks attain their highest densities. It's also where the bog sedges known generically as kobresia grow in thick mats. Here, their broad ground cover protects against erosion, buffers against progressively warming temperatures, and reduces fracturing of the increasingly dry alpine tundra. But changes grow more evident in other areas of the plateau.

On its northern edge, the frozen ground is more than seventy feet lower in elevation than just thirty years ago. One thousand miles to the south, where warming temperatures have been greater along the Indian boundary, permafrost has been lost, dropping the surface two hundred feet lower. The frozen underground is being whisked away by rapid climate forcing. Some of the changes seen in the Arctic, where the floors of freeze-and-thaw lakes crack, and water then drains out, are paralleled here as freeze-thaw cycles shift and as ground temperatures increase, which leads to more cracking.

While some wildlife relies on locally adapted plants like kobresia, other species, such as blue sheep, do not. They prefer grasses, but especially like small, nongrasses called forbs. While these are extensive, blue sheep themselves are harder to find. With an ancestry of sheep mixed with goat, they frequent cliffs where their stocky build and suction-like hooves offer advantages over their sit-and-wait predators.

It's December 5. Thirty sheep, thick in bluish-gray coats, descend a rock-littered slope that is still draped in the long remnants of summer's snow. Aili and I focus our binoculars. One sheep in the group pauses, looks quickly about, and angles downward. Others follow nervously. The animals suddenly panic. Some bolt upslope, away from the snowfields. Others go in the opposite direction.

"Did they see us Joel? Was it us?"

"No, no," I say, and immediately search the cratered landscape for anything unusual, something that might have frightened them. Among the shadows in pitted debris is a wolf. We wait, hoping for more action. Its tail hangs low, curved, which indicates it may be a subordinate.

Strutting into the open, the culprit adopts its true form. Sun illuminates a lush tail. Its fur radiates spots. This is no wolf: it's a snow leopard and the first ever seen by Aili. Me too. The day is special for another reason: it's her birthday.

We try to get closer to the leopard. Scurrying across nearly a kilometer at sixteen thousand feet is not a good idea. I'm left wheezing. We discover a nearby cave, hoping to discover that it's a lair to which the snow leopard has dragged his or her kills and where it has lived for months, but there is nothing inside except one sad set of feces.

It's nearly four in the afternoon. Baima, the fifty-one-year-old policeman and my designated protector, will be upset if we're late returning to the point where we'd been dropped off. I had assured him I'd be fine in the mountains. He didn't believe me. What could an American know about surviving in China's wilds? But with Aili's insistence, we wandered off to explore. Now we're a good few hours away. Baima will dwell on his future in government if I'm injured.

We opt for a shorter route than the one we took to get here, one that crosses a pass nearly a thousand feet higher but then drops into Baima's lap at the Land Cruiser. Having earlier been slow but not too uncomfortable at these mountain passes, I elect for maximum exertion (translation: running). Big mistake. I can't catch my breath. I lie down, seriously gasping. It's a few minutes more until I realize my uncontrolled diaphragmatic throbbing is adaptive. I'll not die, at least today.

I plod on, more slowly. Ten minutes pass. A distant shape moves quickly among boulders: a chain-smoking cop, running at these elevations. Baima is looking for us. Tibetan genetics score an indisputable victory.

Our sampling of wild yaks and chirus continues as the days pass by. Aili has coordinated three teams, all operating at the same time across the Tibetan Autonomous Region so that we can census populations without worrying that we are recounting the same individuals. This is a valid concern since chiru, much like caribou or even pronghorn, migrate long distances. If we count chiru numbers in one area and then a month later in a different region, the same animals might be tallied. Sampling with her design reduces chances for possible duplication. Our minimum counts will serve as just that—the actual numbers seen.

Our surveys reveal more than ungulates. One day a pair of wolves crosses a frozen lake. A different day, one approaches and sits, peering curiously at us from seventy-five yards away. On another day, in golden light, two more appear and stare at us. Either they are very hungry and are trading fear for hope of a scrap or they're unafraid. In these Buddhist highlands, Buddhist persecution of animals is rare.

We see fewer wolves, though, than feral dogs. The dogs are afraid and keep a safe distance. Most are of the Tibetan mastiff variety, a breed unique to these elevations. Herders usually have one or several tied up. They are fiercely loyal to families and fiercely menacing to those who approach. We see more than a few ranging freely. A group of eight works together and chases two gazelles. Another time, we watch different pairs more than sixty miles apart harass kiangs. Elsewhere, others hopelessly chase chirus and, in desperation, Tibetan gazelles. Dogs that kill livestock soon become dead dogs, so it may be that the ones we see are pursuers of native game. Schaller reported more than twenty chirus, weakened by snow, were killed by dogs. Herders are mum on the topic, probably because of the perception that either Baima or the two foreigners—one Han, one American—will finger point about dogs killing wildlife.

We continue ahead, encountering wild male yaks above sixteen thousand feet. A solitary yak crosses cobbles and drinks, unafraid of us, at an unfrozen lead. Another is nestled into a rocky canyon.

Fifteen miles farther, a group of six leaves a frozen shoreline. They sense us when we are a kilometer away and flee. Later, we camp.

My toes are cold. I slip into one sleeping bag tucked inside another. I'm wearing my hat, long undies top to bottom, and loose-fitting socks. Batteries are in my bag, along with cameras and a water bottle. I tuck my jacket under my head and click off my headlamp. The wind rushes. At 3 A.M., my toes hurt. I tear off socks, curl my legs, and cup my feet squeezing gently. Maybe my "loose-fitting" socks were actually too tight. Or maybe the elevation has slowed blood flow to my extremities. Maybe it's just old age. In any case, this isn't good. By morning, the wind has slowed. It's −14°F.

Gradually we enter less remote regions. Clusters of buildings become more frequent. Fences begin to appear. A kiang hangs dead from one, tangled by its hoof, its migration path cut off. Farther down the line is a chiru, also mangled, its body partially consumed by dogs and vultures. In 1903, Cecil Rawling offered his thoughts about the area: "We reached Phobrang, the last civilized spot we should see for months . . . at an altitude of nearly 14,500 feet. It's inhabited all the year round but the people must experience the most bitter cold." Today, the highest communities are located at heights of more than seventeen thousand feet.

Aili and I count over twenty thousand chirus; the other teams—working more remote regions—add another thirty thousand. There were more than five hundred yaks, and almost five thousand kiangs. Numbers are rebounding from the post-poaching periods following Schaller's surveys.

Back in Lhasa, we check into Tibetan Autonomous Region headquarters and prepare for an official dinner. As an honored guest, I'll be seated next to the director of the region's forestry program, an appointee directly from Beijing who's been sympathetic to broad conservation policies and my activities.

Formality is important. Seniority and organizational standing dictate protocol. I desperately need guidance about Chinese customs, so

Aili tutors me. There will be a rotating tray at the center of a table—a lazy Susan. I must be adept with chopsticks, witty, and sincere with toasts. Baijiu, an alcohol distilled from grain, will flow freely, as will chaang, an alcoholic rice brew.

We're led past a few dining patrons in the first room into an ornate, smoky, and windowless chamber. The normal chill of these unheated rooms dissipates in the warmth of a roaring fire. The first course appears: fish head soup, which is followed by a pickled egg dish. Crustaceans come next, and then something with unidentifiable legs.

"You are from Montana?" asks the director, who freely converses with me.

I nod.

"You must own guns. Most Americans own guns."

I'm unprepared for this kind of candor.

"America is a dangerous place, isn't it?"

Aili need not translate: his English is excellent.

A challenge to my cherished citizenry waits while I search quickly but carefully for words, as my response may dictate my future here. I don't take the bait.

"I don't hunt, if that's what you're asking."

Pundits back home would claim I'm a sellout. I have a goal in mind, however: I wish to continue my work here.

Six other attendees, including two deputy ministers and two directorate supervisors, halt their chatter to listen. I've had but one beer; others are on their second or third. More baijiu arrives.

"No," the director responds, "there are many deaths in America because of guns."

"Oh yes—you are right. I think the U.S. has more than twenty thousand murders every year. Most are by guns. One can go to a store and find twenty magazines about guns and shooting but only a few about the environment or wildlife. America can be an odd place."

Perhaps my response will reinforce the senselessness of violence in a country gone astray thanks to gun lobbyists, despite our personal freedoms. "It is sad, and our laws have not allowed what you have done here in China—seize arms to reduce the killing of wildlife."

A smirk slowly rises, turning upward the lips of my gracious host, as if I've been undone. I'm fine with this beginning. I had never intended to discuss Chinese politics or human rights. Why would I? President Obama had just done so nationally on Chinese television during a recent visit, singling out Taiwan and Tibet. I managed to watch this on a fuzzy screen in a remote Tibetan tavern. Some of the Han audience there had watched Obama earlier. Most now watch me.

Later, I steer the conversation to conservation outputs:

> China and the Tibetan Forestry Bureau have made great progress toward reducing poaching. You have not lost a single species. Right? And, you have signed the same legislation as all governments who belong to CITES [Convention on International Trade of Endangered Species] and believe in protection. So, how do you go about keeping species and assuring them a future when others in your government encourage people to settle in protected lands? When herders or resettled people want to chase kiang or wild yaks from their pastures, how do you encourage species in areas that you have said you will protect?

Perhaps now the director will think I'm less of a milquetoast.

"The species matter, and we will do what we need to give them a future," he responds.

Everyone agrees that people and climate are the twin-headed agents of global change and unerring challenges to China's biodiversity.

Wild yaks have dealt with wolves for millennia. Perhaps they encountered high-elevation tigers, which still occur up to more than eleven thousand feet. They probably dealt with cold-adapted hyenas; the steppe variety overlapped Asian bison. Wild yaks met other hunters, especially humans, who have been on the plateau for more than twenty thousand years. Their antipredator senses are keen, and some accounts suggest they faced their enemies in groups. "When in danger they form a phalanx, the calves in the centre, some of the full-grown males advancing to reconnoitre," observed Nikolai Przhevalsky in the 1870s.

In 1900, Cecil Rawlins wrote: "When a herd obtains the scents of a human being, they all rush together and remain thus with their heads toward the threatened danger." I again think of muskoxen.

But Rawlins also noted that they "move away at a trot and a gallop, their tails waving from side to side keeping this pace for hours and never halting until many miles of country have been covered." Wild yaks obviously share responses like their North American equivalents—bison—for which the greatest single recorded flight exceeded fifty miles in twenty-four hours. That was in Canada's Wood Buffalo National Park.

Less is known about relationships between wild yaks and wild carnivores than those between carnivores and northern bison. In Tibet, I once watched a solitary wolf approach within ten yards of two males before they took notice. Another time, a yearling yak was driven onto a frozen lake where it was killed by wolves. And Schaller once watched five wolves pursue a group with calves. All disappeared into the hills, and the outcome was unknown. Although we hardly witnessed interactions between yaks and wolves and none with bears, it's possible to make some simple inferences about how predation affects the yaks' use of their high-elevation habitats.

Gender and behavior dictate reproduction. While the effects of evolution are not ordinarily very random, knowing something about the two sexes across time enables a level of deduction, something Darwin knew well: "It is rendered possible for the two sexes to be modified through natural selection in relation to different habitats of life, as is sometimes the case; or for one sex to be modified in relation to the other sex, as commonly occurs." Darwin's prescience had as much to do with mammals as other groups where, for instance, the likes of male and female peacocks, elephant seals, or beetles differ so conspicuously in body size, color, or ornaments that they might be mistaken for members of another species. Such variation in size and sociality undoubtedly affects how the sexes live.

Males typically experience greater mortality and live shorter lives than females, a pattern that extends from fruit flies to humans. The reason is primarily testosterone, which in turn governs the zest by which males behave and their inclination to aggressively acquire mates. But females just don't idly wait. They, too, command their fates, sometimes actively selecting mates while at other times creating safe zones to raise their offspring. It's this part—where to find food and where to protect babies—that enables insights about how predation may have shaped land use by yaks.

More than a century ago, British explorers indicated that wild yaks lived in "the coldest, wildest, and most desolate [treeless] mountains" and could be found "northwards to the Kuenlun, at elevations between 14,000 and 20,000 feet." In 1892, Alexander Kinloch succinctly described groupings: "The cows are generally to be found in herds varying in numbers from ten to one hundred, while the old bulls are for the most part solitary or in small parties of three or four." Even earlier, in 1875, the American naturalist Joel Allen commented on the propensity of male and female bison to live separately, mentioning that the "habitual separation of the large [female] herd into numerous smaller herds seems to be an instinctive act."

That sexual segregation occurs is indisputable. The question of why is a different issue. Food, fear, or something else? Perhaps answers are as simple as knowing what the sexes eat or where they die. Or the issues might be more complex and require understanding the degree of habitat choice, effects of variation in male and female size, or a preference for groups that are large or small. Maybe answers will become apparent when we know how choices change as individuals grow older or how prior experience shapes an animal's psyche—even whether past persecution molds today's choices.

Wild yaks carry a hidden story. It's shrouded in a high-elevation enigma that may change as Asia's lofty heights deglaciate. They wear no radio collars and never have; such tracking devices are forbidden in militaristically sensitive and remote regions. We thus have limited tools for assessing where wild yaks occur. Such tools as we have mainly include our meager ground observations and past accounts. The other tool is inference.

Like many ungulates of extreme topographies, males generally travel to higher elevations than females. Some in Alaska's mountains do just the opposite: female caribou reach higher elevations, at least in June during the birthing season when their energetic needs are greatest yet food is scanty. Terry Bowyer, an Alaskan ecologist, collected data on food quality and predator abundance and ultimately concluded that female caribou exchanged more plentiful foraging at lower elevations in order to avoid wolves and grizzly bears.

Biology, like so much else, involves trade-offs—the costs of some activity balanced against the benefits of another. For caribou, it's better to have your baby alive and incur the high costs of low-quality food than to access high protein at lower elevation but risk costly interactions with bears or wolves. In the end, if your baby is mere veal, then your investment is for naught. Perfect solutions do not exist. Trade-offs do.

Among wild yaks, both males and females live just below glaciers, at average elevations between 15,700 and 15,900 feet in winter. Group sizes differ dramatically. Those groups with cows typically number thirty or so, although assemblages can be in the hundreds. For males, it's the opposite; if they are not solitary, they live in tiny groups—with usually fewer than three individuals and rarely with more than ten. On an absolute scale, males go to higher elevations but not consistently. The highest we ever found one was at over 17,200 feet, though others report they've spotted wild yaks at elevations of nearly twenty thousand feet in summer.

Like yaks, bison males also reach greater heights than females. In the Rocky Mountains, contemporary skulls have been found at 12,000 feet in Wyoming, 12,600 feet in Utah, and above 12,750 in Colorado; they sometimes emerge from melting glaciers. The highest elevation at which a group of females was found to cluster was at 8,350 feet, but these were old—of late Pleistocene origin.

Is it relevant to know which sex goes higher or is this just an idle factoid? Given increasing temperatures, there will now be little recourse for animals other than to move up, once the mountain's highest meadows and permafrost shift to trees. Suitable habitat will not remain. Nevertheless, knowing how current habitats are used has value. In the case of yaks, with increasing numbers of domestic yaks, goats, and sheep and people, changes in distribution will assuredly occur. For instance, during our surveys, the number of male wild yaks on steppes and arid grasslands was three times more than females, who instead alighted on slopes and more precipitous terrain.

Where this gets more intriguing is that the elevations of female groups with calves and those without did not differ significantly. The difference was basically between 15,825 feet and 16,035, which is biologically trivial. Females with calves were in the hills or on steep slopes but absent from the flatter steppes. Those groups lacking calves were out in the valley bottoms on two-thirds of our observations.

How might we interpret the variation in habitat use? One possibility is that the differences in groups with and without calves is just random, a proposition we tested statistically and found so unlikely that it would occur in less than one in fifty trials. Something else must be going on. Another possibility is that yak females with calves do things differently, just like the maternal caribou used high elevation to avoid predators. Lactating bison of South Dakota's Badlands actually use open areas with better visibility, whereas females without babies more often stay in ravines and gullies.

In 1876, Przhevalsky was far ahead of the curve when considering trade-offs and predation: "Cows and young bulls are . . . exceedingly difficult to kill for another reason, viz., that they are always in troops, and you cannot single out one out of a herd. . . . They are also more wary and difficult to stalk than solitary bulls."

Predation alone cannot, however, be the sole explanation for choices by wild females with calves, because they certainly derive nutritional benefits through their primary selection of steeper post-glacial topography with its kobresia, streams, snowmelt, and slopes. Again the argument relies on inference. We know from domestic yaks that mothers consume sedges and follow a high-elevation gradient into lofty alpine meadows; these areas, like the slopes preferred by yaks, retain protein longer than regions lacking in water or snow.

Heinrich Harrer, author of *Seven Years in Tibet*, was once a mountaineering partner of George Schaller. Heinrich knew domestic yaks for what they are: snow plows. As sure-footed transporters of supplies and providers of sustenance for herders, entire agricultural institutes are dedicated to the study of yaks. This is a sensible investment from which to understand human livelihoods. By contrast, we know little about their endangered, wild progenitors. We don't know how many there are or whether predators are serious threats. We don't even know how long females lactate.

We push off in our search for more wild yaks. The Purog Kangri Glacier comes into view and then fades behind wind-whipped clouds. The glaciated massif reaches nearly twenty-three thousand feet. We spot little wildlife. The next day we approach a salt lake. Beyond stony shores, a meadow is thick in sedge and ice. Alongside the lake are five yaks in violent motion, tails high, and engulfed in dust. We assume they're wild. Existing dogma is that wild yaks run from people. Domestic yaks do not.

Wait: Does running define them as wild, and standing as tame? How can that be? That's circular—a tautology. Is there no independent metric to judge wild from domestic? I peer quickly through my binoculars. A dab of white is on one's tail. Domestic, I believe. They disappear over a ledge. My glimpse is fleeting at best.

An 1890 report summed up the value of the tails: "The bushy tail of the yak is well known, being highly prized in India . . . for switching away flies, and [they] used to be considered as emblems of royalty. The white tails which are brought for sale, are those of tame yaks; tails of the wild species are black and of much greater size." Typically, the solid black tail of the wild species has rusty brown overtones, and in the far west of their range, there is a subpopulation with golden fur. Now, more than a century after the above report, the distinction between wild and domestic remains fuzzy. Investigations of genetic purity are still lacking, and the taxonomic distinction between wild and domestic yaks remains uncertain and sometimes contentious.

The purity of yaks has an uncomfortable parallel with bison and how we understand and classify each of them. Already something on the order of 99 percent of North America's half million bison contain some form of cattle genes. So do we call such bison wild or domestic? And for yaks in what we call the wild—meaning remote—areas of Tibet, is the best diagnostic tool for distinguishing them from domestics really a denotation of color or flight behavior?

When I asked Shaun Grassel, a PhD-level biologist, for his opinion about the genetic mixing between bison and cattle, his Lakota view reflected that of many Native Americans: "If they act like bison and look like bison, they must be bison." By analogy, if free-roaming (nondomestic) yaks have a peppering of white on their tails or faces, yet group like wild yaks, graze similarly, and segregate sexually like wild yaks, then just maybe they fulfill the same ecological role as the wild species.

It's true that ecology is not genetics, but why are many scientists so stern when it comes to genetic integrity? To find the answer, we need to look backward in time.

Breeders of domestic stock point to the infirmness of hybrids—lighter weights, poorer survival, and, at times, sterility. These would be true hybrids, or the results of crossing different species, like a horse and an ass to make a mule. For yaks, some Tibetan herders prefer the consanguineous matings between *drong* (wild bulls) and *dri* (domestic females), because robust offspring result.

On the one hand, genetic distinctiveness is important because populations may have local adaptations that enable persistence under unique conditions. Undoubtedly this is true. Desert bighorn sheep do better in extreme hot and dry environments than their more powerfully built mountain-dwelling Canadian relatives. But is there really a downside if somehow these northern sheep contain a few genes from southern relatives—called introgression—or from domestic sheep? Or would it matter if the northern sheep had some genes from ancestral sheep in common with a different species, such as the Dall from the Yukon or Alaska? The few studies conducted on introgression, though none on sheep, often report inferior genotypes are whittled away by natural selection.

Exceptions do exist, however. When one sees black wolves one is really seeing dogs, because the dark ones emanate from domestication. What we see in the present is a reflection of past backcrosses between wild and domestic animals gone wild. Natural selection has not removed blackness. While some conservationists suggest we reconstitute the purity of some wild species, perhaps using ancient DNA or by selective removal of individuals with genes of domestics, it's not an argument I hear for black wolves.

Among the most notable cases where genes from domestics have been infused into the wild are those of wild horses. In their native form, these marvelous duns with stubby stiff bristled manes once spread from the deserts and steppes of northern China to Europe. In Mongolia, they're called *takhi*. Elsewhere in the world they're known as Przewalski's horses.

Long thought extinct globally, Nikolai Przhevalsky glimpsed a few in the 1870s north of the Tibetan Plateau. In the waning years of the nineteenth and early in the twentieth centuries, a few *takhi* were captured and showcased in zoos. The last capture of a wild Przewalski's horse was in 1947—a female foal north of Xinjiang, in the Bajtag Bogd Nuruu (*nuruu* means mountains or a range in Mongolian). That was in Mongolia, beyond the purview of today's formal China. Named Orlitza III, she lived until 1973. The coup de grâce for wild Przewalski's horses and options for later captures closed in the 1960s, when the last were seen.

Orlitza III produced four foals with other wild horses, several of which also had young, and these few founders broadened the success of captive breeding efforts. This process was dicey and not just because many Przewalski's horses died during World War II, but also because the genes of a domestic mare were spread among surviving lineages. It was, however, meticulous zoo records of pedigrees that enabled insights into the interplay between genetics and survival in small captive populations. Where levels of inbreeding were high, birth weights were low and offspring survival poor; these facts have reinforced why genetic diversity needs to be part of every species management plan.

What this has to do with yaks is perspective. If pure wild yaks have a better chance of thwarting disease, adapting to changing conditions, withstanding the vagaries of unpredictable weather, or consistently yielding larger babies, then mixing with domestics may not be to their advantage. Conservation geneticists say that evolutionary purity matters. Maybe. Think of wolves and the color black, which exists in wolves only as a result of mixing with dogs.

Does seeing a patch of white in a putatively wild yak matter? I prefer the all-dark forms but I'm not sure why; they seem more wild I suspect, but it's an argument of the mind. Rich Harris reported none of the yaks on the plateau's eastern margin that he presumed were wild had a mix of white. Similarly, less than 1 percent of the ones we saw a few mountain ranges away from that site had a dash of white. A nonrandom set of photos sent to me from the southwestern edge of Chang Tang, nearly six hundred miles west of our study site and closer to herders, shows a couple of huge, all-black bulls (*drong*?) that were mixed with *dri*. About 10 percent of the young had silvery etchings.

The reality is that little is known about the frequency of mixing within yak populations. No one seems to be scooping poop and using DNA to distinguish wild genotypes from those of the less wild. The intensity of selection in wild yaks awaits discovery.

We crest another ridge. Sixteen yaks are below. The group is mostly females, and the slopes are steep. We're at 17,400 feet. They bolt. The snow above looks deep; they run nimbly, including the calves, which weigh a couple of hundred pounds each. I'm still unclear as to why they're so flighty.

Many species possess legacy effects, attributes of a population shaped by interactions over time—just as a sports franchise might have established winning records for decades, like the loathsome Yankees or the converse, Chicago's once winless but now celebrated Cubs.

In biology, legacy effects are common. Some are products of past evolution like eyeless fish in caves, but others are behavioral. Elephants and elk change their migration routes, a behavioral heritage that stretches back to fear of predation and to shifting foods; the animals have come to select passages less onerous than those before. Coyotes learn to fear humans when persecuted. In the Arctic, it's the same story: if you chase and dart muskoxen from a helicopter, and they remember.

What about wild yaks? Are there legacy effects?

I can't answer. We can't use helicopters because the elevation is too high, the air too thin, and the government unwilling.

> Notwithstanding their sterility and the unfavourable conditions of climate, the deserts of Northern Tibet abound with animal life [wild yak, the white-breasted argali, the kuku-yaman, the antelopes called orongo]. . . . Though food is scarce, they have no fear of encountering their worst enemy, man; and far removed from his bloodthirsty pursuit, they live in peace and liberty.
>
> NICOLAI PRZHEVALSKY, *Mongolia, the Tangut Country and the Solitudes of Northern Tibet*, 1876

9

Birthplace of Angry Gods

It's November 23rd, and cold outside my four-star hotel in Nagqu, somewhere near the center of Tibet's Autonomous Region. My bathroom door is bolted shut. That's a problem. It's 5:50 A.M. I need to go, and badly. I grab my headlamp and hobble a hundred yards down an icy street to the decadent public outhouse. At least the human by-products are frozen solid at −1°F.

Stars shimmer. A sliver of moon and Venus glitter in crisp mountain air. Back in the hotel's kitchen, it's a balmy 22°F, the door ajar all night. Two Tibetan women greet me, and know immediately I'm prowling about for boiling water. The only liquid back in my stylish bedroom is the wash bucket coated in ice. Today my coffee will be dark, a French roast that I've brought along. My twin hosts are less enthused about coffee but happy to preheat my thermos. They're also curious about the paper filter that keeps the grounds out of my soon-to-be morning glory.

Only a few weeks earlier, I was in a more modern hotel. I had no worries about electricity or bathrooms. In Beijing, the stars in the sky are few, but the air pollution is rich. The attendants at the front desk spoke English, or good enough anyway. By chance, the preeminent George Schaller was there. It was nearly 10:30 P.M. when I arrived, so I handed the clerk a note and stressed that it was to be put in Schaller's box for the morning. The clerk nodded. Perhaps George would meet me for dinner tomorrow.

At 7 A.M., there was a pounding on my door. Schaller was returning the favor, he explained: at 10:35 the night before, the innkeeper had banged away on his door, saying, "I have a very urgent note for Dr. Schaller." *Urgent?*

That evening George and I sit cross-legged on his floor while sipping wine and unfurling topographic maps. We explore lines that are contoured tightly (mountains) and flared widely (valleys). We redesign surveys, and rethink data needs. With typical graciousness, George offers his prior yak data—specifically those from his and Aili's nine hundred–mile crossing of the plateau's northern edge in 2006. It's part of the area that I'll subsequently work—situated below the Kunlun Mountains and their changing glaciers and extinct calderas. My base will be Kekexili, a vast unpeopled land the size of West Virginia. In Khalkha, the common language of Mongolia, *Kekexili* means attractive young girl. I look forward to the attraction.

We muse about our fascination for old maps, while at the same time acknowledging the power of satellites to interpret landscapes, glacial boundaries, and snow cover. George asks: "What's happened to the love of fieldwork in modern science? Does natural history fit anymore?" Two years will pass before I will again see the maestro.

On another trip, it's December at our base, a little above sixteen thousand feet. I'm loath to leave my two sleeping bags, one snuggly tucked inside the other. My slightest motion triggers an avalanche of the hoar frost coating the inside of my tent. The temperature is −24°F. I escape my toasty nest and walk to our walled canvas tent. One driver uses a gas-propelled torch to melt ice blocks so we have drinking water. The other, already outside on his back, aims a blue flame at the crankcase of our four-wheel drive vehicle. Every few hours throughout the night they do this to keep the engine blocks from freezing.

Our camp is in the shadow of the Bukubada Glacier, a massive vault covering 170 square miles. To our east and north are smaller glaciers—the Yuxu reaching 19,619 feet and the taller Yuzhu at 20,500 feet. Then there is the big one—Muztag, 25,341 feet, to our west. All converge near the juncture of Xinjiang, Qinghai, and Tibetan Autonomous provinces.

We're looking for yaks in the immensity of Kekexili National Nature Reserve. To receive permits, I had agreed not to chart routes with GPS, a demand of state authorities. We're never told the reason. At their request, however, we were to map the locales of kiangs, yaks, chirus, blue sheep, and bear. Gazelles were so abundant they were of no interest, and neither were wolves. But, by plotting the wildlife sightings the government officials are interested in, we would still be able to re-create most of our route because sightings would obviously all be sequentially logged at the specific locales recorded. These data, in turn, could be useful for estimating densities and deriving a first approximation of relationships that involve influences of glaciers and weather on animal distributions.

Earlier travelers well appreciated such complexity. Writing in 1854, Alexander Cunningham described the region, saying: "This climate is equally favourable to animal life. The plains between 16,000 and 17,000 feet are covered with wild horses and hares and immense flocks of domestic goats and sheep." Thirty-five years later, in *Through Unknown Tibet*, Montagu Wellby commented: "We crossed some hills, and then descended into a valley of sandy soil. There was also rich grass, and several small streams flowed across the valley, taking their rise from the snow peaks north of us. Here we saw immense herds of antelope, all females and young ones . . . thousands and thousands."

Finally, in 1906, the connection was made between liquid and life, through the eyes of Graham Sandberg: "The annual renewal . . . to counter-balance in some degree glacial waste on the higher mountains is an appreciable factor. . . . Every atom is preserved and goes to increase those stores which replenish glaciers, and through glaciers—in due process—the resultant rivers."

To our east are nomads and herders, and China's mother of all rivers, the Yellow—or in Tibetan, the Madoi. Had Sandberg lived until 1997, he'd have been surprised to learn that the Yellow had dried up and disappeared for almost four hundred miles. The situation triggered serious alarm among Qinghai's government officials and generated massive interest in the health of these high-elevation systems, both for water retention capacity and the grasslands that support agro-pastoralists.

The Tibet-Qinghai Plateau's ecological dual persona of aridness and high elevations means that the region's fragile ecosystems are especially susceptible to even subtle changes in temperature and rainfall. Increased run-off, changing lake levels, and potential long-term impacts of water flow from the headwaters of critical rivers such as the Mekong, Indus, and Brahmaputra are unknown. Even short-term effects are not well understood, though widespread environmental and economic impacts are increasingly felt in the enormous downriver watersheds. Higher up, the warming and increasingly variable precipitation exacerbates conflicts between humans and wildlife. This discord is seen by way of reduced grassland productivity, which gets amplified when species like kiangs flock to pastures used by herders. With reduced forage, the yield of domestic yak meat and milk is compromised. Reproduction in gazelles, antelopes, wild yaks, and other species will also be affected. Perhaps the birthplace of angry gods is just too remote for the world to seriously consider that issues there are similar to those many others also confront.

In America, the resonance of glacial loss gains momentum from the Crown of the Continent, a wonderfully untrammeled protected realm in the northern Rockies where less snowy peaks dot the splendor of Montana's Glacier National Park. There, mountain goats occupy the alpine's semi-permafrost, and caribou once browsed the hanging lichens of old growth conifers at lower elevations. When designated as a park in 1916, Glacier had a hundred fifty ice sheets; today, 85 percent are gone. There are now three times as many days annually with temperatures above 90°F. Streams are warmer. Bull trout, reliant on the cool water, are retreating to higher elevation streams, yet still struggle to persist. Ski resorts have closed. We don't yet know about what the effects will be on goats, whose nearest Tibetan affinity are blue sheep, but all continents are being pinched by temperature regime changes. I want to better understand the twin demands of climate forcing and humans on wild yaks.

Journals from the earliest Western explorers of Tibet hold clues. The first was written by George Bogle in 1774 but his descriptions were vague. Those of others, however, recorded observations of yaks by habitat and by grouping. Between 1819 and 1825, William Moorcroft and George Trebeck did so while journeying from Ladakh and Kashmir to Bokhara in modern day Uzbekistan. Twenty years later, Godfrey Vigne described wild yaks, although with insufficient detail for my use. Fifty-seven additional accounts, however, by Russian, French, Swedish, British, American, and German expeditions yielded usable annotations, which amounted to twenty-eight trips with 217 inferable observations of male and female yaks by habitat. I assigned these either to steppes and valleys, or to hills and slopes, and narrowed the sightings to 148 operational data points for more rigorous analyses. In 1876, for instance, Przhevalsky noted it was on "the southern slope of this range that we saw herds of them." He goes on to say: "I was clambering over the mountains when I suddenly caught sight of three yaks lying down." In 1898, Wellby wrote: "I spotted in a valley . . . a single yak grazing," and Cecil Rawlins, in 1905, recorded that "immense herds of wild yak were to be seen grazing in all the valleys."

Overall, twenty-five of the historical surveys contributed 65 percent of the observational data, while Rawlins, Wellby, and Henry

Deasy contributed the remaining 35 percent. Expeditions were primarily during winter because summer travel was marred by inefficient crossing of rivers and marsh.

During the period of exploration and discovery late into the nineteenth century, both males in their small groups and females in their much larger ones were found at about the same rate on flat areas, valleys, and steppes. This was before major poaching epidemics began. Part of the appeal of the Kekexili region for me was not just that Schaller had surveyed there, but that we would also track some of Wellby's 1890s route.

When planning that journey while yet stateside, I realized I would need someone to accompany my all-Chinese team and me into the remote Kunlun Range on the northern side of the plateau—someone who speaks familiar English, who understands the formalities of research design, and who doesn't whine. In short, I need someone who relishes field biology. Other attractive traits could include experience in Asia, indifference to cold, willingness to work long days, and familiarity with science. Although there are many good candidates, my permits have yet to clear, and that's the issue. Most people can't just drop everything on a whim. But one person comes to mind: Ellen Cheng.

With a PhD in ecology, experience working with Canadian fur trappers, a fluency in molecular genetics, and a lifestyle that differs from most Americans, this petit Floridian by birth was somewhere in Asia. Previously a Fulbright scholar in Ulaanbaatar, Mongolia's booming capital, Ellen was now tucked away in the Himalayas. Living in Lamai Gompa, beyond a town called Bumthang, she advised Bhutanese scientists on field studies, helped design sampling schemes on species as different as tigers and amphibians, and was the architect for the emerging Bhutan Ecological Society. My emails to her go unanswered, maybe due to power outages, maybe because she's avoiding contact, or maybe Ellen is just gone. Regardless, I'm out of luck.

It's Thursday. The Chinese Embassy in Washington, DC, has yet to return my passport. Without it and the visa, there is no travel to Kekexili. This is getting dicey. My flight departs Missoula Sunday morning. For the fifth time I check emails. The embassy is as silent as Ellen.

There is a flicker of hope. The power is again on in Lamai Gompa. Ellen Cheng expresses interest in joining the team, but the logistics are formidable.

It's already Friday in Bhutan; Ellen lives twelve hours from Thimphu, the capital. Freeways have yet to show their asphalt faces in the Buddhist kingdom of stupas and stunning scenery. Earthen roads over precarious mountain passes remain slow and close often. Given that avalanches and rockslides are typical events, the Bhutanese and Cheng accept slow-paced lifestyles. In a country where the gross national product is happiness, there is more to life than timely arrivals.

It's now Saturday. My three duffel bags are filled with cold weather gear as I'm prepped to tackle late November and December in eastern Tibet. Still, the passport and visa have not arrived.

The unrelenting Ellen is determined to make things work on behalf of the yak project. Her plan is to fly to Thailand and then camp out at the Chinese Embassy there since Bhutan has no legation and an in-person appearance is required by the Chinese Embassy in Bangkok before she will be allowed into China. I don't know whether her being Taiwanese-American will help or hurt her chances.

At 4:30 in the morning in Bhutan, she hoists a sixty-pound pack onto her back and walks in darkness and cold at seven thousand feet. The bus stop is just a three-mile stroll from her base in a walled government compound. From there, it's a twelve-hour bus trip to the capital, and then a real hustle for transport to the airport in Paro. If there are no flat tires or road closures, and if buses are timely, she'll make the flight to Bangkok.

Back in Montana, my own problems grow more serious. It's 5 P.M. Saturday, less than twelve hours until I'm scheduled to depart, and I still have no passport. The FedEx website indicates not only that my passport was sent but that it was actually delivered to my home. *My home! Whaaat?* I'm at my front door, on the phone with Fed Ex. They swear they delivered it. "Believe me, there is no package here," I tell them.

They'll recheck—lovely.

Wait. A man with a little kid in tow walks down the street toward my house. In his other hand is a large white envelope. He lives two blocks away, and I don't know him, so I'm especially grateful that he didn't wait until Monday to bring the misdelivered envelope to me. Sunday morning, I board for Beijing.

Back in Bangkok, Ellen is interrogated at the Chinese Embassy there. But two days later she, too, is in Beijing. So is Lishu Li, a Chinese national now working for WCS on illegal trafficking of wildlife. In a prior life, she was a grad student of mine in Montana.

The three of us manage to book a flight to Xining, the capital of Qinghai (see map, chapter 7). The layover is thirty hours, and so we go to the zoo, meet ministry officials for permits to be in restricted areas, and visit a Buddhist monastery thick with police. While we are wandering the hectic business district, it begins to snow. A man lies naked and curled in a fetal position, his body scorched by fire, an immolation we presume. A lone piece of cardboard substitutes for clothing. He's barely breathing, but his eyes are open. People stream past. Shocked at the indifference, Ellen and I ignore overhead cameras and Lishu's fearful plea to walk on. We stop, leave some money, and then move along. The immolations continue.

The next day, we embark on our nine-hour train ride westward—420 miles—to Geermu, also known as Golmud, a Mongolian word for rivers. There, at an altitude of ninety-three hundred feet on the eastern edge of the great plateau, we grab meat and milk, drums of petrol, and spare parts, then head out in two jeeps. Lishu's fresh vegetables freeze almost immediately.

The search for yaks takes us through unpeopled valleys. The land grows high and silent. We follow a river gorge sculpted by a long-gone glacier and travel through narrow canyons. Crossing a mountain pass at fifteen thousand feet, I'm reminded of a decade-old observation of golden-colored dholes attacking blue sheep. While these pack-hunting wild dogs are nowhere to be seen at the moment, wildlife does start to appear. First Tibetan gazelles, and then, in the gentle light of morning, kiangs. Big-game hunter Alexander Kinloch wrote of their behavior in 1892: "In places where they have not been disturbed[,] Kyang will frequently gaze at the sportsman within fifty yards, without any fear, but merely curiosity. On the more frequented routes which are annually traversed by tourists the Kyang are much more shy, and seem to know the range of a rifle well." We wonder whether, within the confines of Kekexili's protective border, these large equids differ in behavior from those on the outside. The ones we watch whirl and bolt in astonishing synchrony below shimmering glaciers.

As we watch, wolves, the ecological counterpart of dholes, appear. One shows particular interest in us. I crawl toward it. Caught somewhere between curiosity and wariness, the thickly furred lupus approaches to within ten meters of me. It edges closer. I stand, and slowly back away.

Thirty miles farther, in a valley rimmed by mountains pushing eighteen thousand feet, chirus prep for the rut—sometimes in groups of a hundred or more. Intent on intimidating and likewise wooing females, males boldly display their rapier horns. Bluffing is common, and fights rare.

While we pass the days looking for wildlife, our nights are spent trying to stay warm. Sometimes we're in tents. Other times we use unoccupied mud-and-brick huts as wind blocks. As we heat rusted stoves with sheep or yak dung, choking fumes seep through, leaving a smoky haze. My morning coffee seems to assuage the respiratory jolts engendered by the fumes. Then, we push on.

At times, lakes block our paths. Though the lakes are frozen, we deliberate about whether they will provide safe passage. Some stretch for fifteen miles along shorelines of scree and loess, the remnants of recent glacial loss. Detouring along the rocky beaches will require almost half a day and precious petrol for the jeeps. But the ice has risks of its own: a month earlier, three geologists died when their vehicle broke through in a failed crossing.

In the 1840s, Évariste Régis Huc, Joseph Gabet, and others traveled the plateau and "observed dark and shapeless masses, . . . fifty wild cattle, absolutely encrusted in ice." They wrote: "Their fine heads, surmounted with great horns, were still above the surface; the rest of thc bodies enclosed by the ice." I'm reminded of the fifty-two muskoxen frozen we discovered in the Arctic *ivu*. In 1892, Gabriel Bonvalot described another tragedy: "A little farther on we find emerging from the ice the almost intact humps of camels and upon closer examination we see that part of a caravan has been drowned here, including the camel-driver, one of whose arms is raised. . . . This must have happened only a short time ago, when the ice was not thick enough to bear them. We have nothing of the kind to fear, for the minimum of the night was 18° below zero."

With such cautionary stories in mind, we stop our two vehicles and wonder whether the lake's ice crust will bear weight. At first we cross on foot—only a few pings and some slight fractures, all good. The first vehicle sets out. As it touches shore ice, a fissured vent shoots forth, cracking for hundreds of feet. We're undaunted but cautious: we unload the vehicles, rolling fifty-five gallon drums across the surface to the opposite shore, along with supplies. With the trucks emptied, we gingerly set sail.

Days later, we reach Zhuonai Co (*co* is lake in Mandarin), another vast water body. Because it's half saline and so freezes at a lower temperature, we're unwilling to risk that crossing. After camping for the night, we'll detour around. We think of what this place will look like come June; the adjacent steppe will be the lifeline for thousands of female chirus as they complete their migratory journeys. Also becoming more prominent will be eagles and wolves, attracted by the synchronous pulse of spindly new calves. Brown bears will also notice those calves.

In *The Land of the Lamas* (1891), American diplomat William Rockhill described his encounters with bears. "I was riding ahead of my party when I saw in a little pool a big brown bear eating a dead yak. . . . Bears are very numerous around the Yellow River where they do not keep to the hillsides, but are frequently met with on the plains. Holes about five deep and as many broad, dug by them, were continually passed; the Mongols said that these were their dens, but it is more than probable that they were dug by them when hunting for their favorite food, the lagomys [pika family]," which live in burrows.

Rockhill was incisive: more than a century later, surveys in the Zhuonai region found that nearly 50 percent of the brown bear diet was pika, 30 percent yak, and about 15 percent chiru. A couple of thousand kilometers away in the Chang Tang, Schaller reported that 60 percent of their diet was pika. Whether bears in Kekexili were predators of ungulates or opportunistic feeders of carcasses isn't known. Nor is anything known about relationships between wild yaks and wolves. These fearless wolves, in particular, are worthy of a future project. I'll try to find funding for work on predator-prey dynamics.

Over the next week, we see more yaks, with the males generally out on the steppes and females tucked away in the hills. In the mornings, we scan the panorama, drive, stop, scan, and hike. We cross frozen rivers. There is little available water. While the cold undoubtedly reduces individual needs for fluids, all mammals require moisture. Snow can be substituted, but ice does not do the trick. Pastoralists have known this for millennia, as their stock suffers if only ice is available. In the 1890s, Bonvalot was aware of this as well: "Our arrival puts to flight a dozen orongos [chiru] licking the surface of the ice, which shone in the sun like a mirror."

One day, we find a group of yaks below glaciers. It's a large assembly, and even with our scopes we can't get a good estimate. So Ellen, Lishu, and I dress warmly—face masks and thick gloves—hoist on packs, and pretend we're alpinists in order to approach for a decent count. The wind off the Bukubada Glacier is intense. Two hours later, higher, and totally frozen, we erect tripods and scopes. They're unsteady in the gale. We push on. As we step over a ridge, the yaks see us. The lead cow walks and doesn't run. She pauses and with seeming foresight leads the herd across rubble. We manage a count—thirty-three, less than we suspected.

Another day, we glimpse a group crossing a suture between steep ridges at seventeen thousand feet. Later, we find more down low. Curiously there are no young. Further ahead is a large congregation—some two hundred lulling about loamy soil near patches of sublimating snow, yet it is a puny gathering in relation to historical observations.

Days pass by. On one we might observe a solitary male moving through a narrow valley. On another, two *drong* lie in shade on a −15°F morning, unfazed by our passing.

A different time, three bulls charge as we sit in our jeep. With tails up and hot-breathed ferocity, they send loess exploding—a scene that's evocative of the same story line but in far different locales: desert, aridity, and a large behemoth coming fast. It could be rhinos, but it's not. This desert is cold. Had our intent been food, the trio would be easy meat. We accelerate away.

Although past accounts vacillate between describing them as docile and describing them as cautious, these yaks—at least the males—do not appear fearful. Perhaps that's because they see few people or because they are not persecuted. Rockhill noted that "the hills around this plain, and also Karma-t'ang, were literally black with yak; they could be seen by thousands, and so little molested by man have they been, that we rode up to within two hundred yards of them without causing them any fear." This was soon to change.

Starting in the 1930s, serious killing of yaks began. It reached its apogee in the 1980s and 1990s. Unlike the slaughters of American bison in the nineteenth century that were associated with building railroads and resulted in killing off the food supply of Native Americans, the poaching in western China was done by highway workers, illegal gold miners, the military, and the police.

Today, much of the poaching is under control. At the eastern edge of Kekexili, we pass an ornate memorial to venerate two rangers who died as a result of their protection efforts. Along the seven hundred–mile paved road that connects Geermu to Lhasa, cars and trucks whiz past. Eighteen-day summer bicycle tours are advertised. Travelers along this highway might glimpse wild yaks, kiangs, and even chirus. Qinghai vice governor Ma Peihua praised the chiru, which, he said, "embodies the Olympic spirit of higher, faster, and stronger." And yet the mascot for the 2008 World Olympics was not China's most renowned citizen, the panda. It was, instead, the more powerful chiru.

The U.S. capital rarely has a slow day, but January 21, 2013, was one. The headline in the science section of the *Washington Post* read: "Yaks, Listed as 'Vulnerable,' Are Making a Comeback in Tibet." Not only did the article offer a snapshot of the 991 yaks we counted, but it carried photos of our expedition and communicated our optimism about the yaks' recovery. Yet nature's handiwork is never finished and protection never guaranteed. Eleven months later, the Xinhua News Agency reported a winter poaching event. In the Xinjiang Uyghur Autonomous Region to the northwest of Kekexili, authorities apprehended poachers with seventeen fresh carcasses in their truck. These included eight yaks, six chirus, and a lynx. All are under first-class, state-level protection. A frozen river was the conduit to access the reserve.

Poachers and police, and predators and prey, all play in the same biological drama of detection, capture, and avoidance. Whereas poachers try to minimize discovery to avoid incarceration, prey and predators are together playing a more life-and-death game, one that varies across ecological zones and deep evolutionary time. A century ago, during a period predating pervasive poaching, male and female yaks used steppes and valleys at about the same frequencies. I wondered whether those past occurrences of habitat use had changed in the relatively better protected confines of today's Kekexili. If so, is current habitat use a legacy effect from past predation, whether by carnivores who hunted for food or armed humans who insatiably slaughtered?

Harassment is the key factor here. Historical reports from Tibet suggest that, where unmolested, yaks and chirus had little apparent fear of humans. Intrepidness can be ubiquitous if animals are not harassed. In East Africa, Serengeti lions barely respond to humans but are much shier in areas where poaching occurs. In the Altiplano of the high Andes, the behavior of wild camels (guanacos and vicunas) reveals the recent history of their harvesting, legal or otherwise. In North America, grizzly bears with only a modern history of being killed by rifle-bearing humans are aggressive toward people, while bears in Europe, where serious persecution has occurred for centuries, have learned to simply avoid humans.

Even Arctic whalers noted legacy effects. In the 1880s, John Murdoch wrote of Barrow, Alaska: "Not only have the walruses been killed off by the indiscriminate slaughter which has been the custom, but they have grown cautious, and have learned to withdraw to inaccessible parts of the ice fields, where they cannot be reached with a boat." I had a sense of déjà vu with Kinloch's 1890s account of kiangs that had learned to remain beyond rifle range.

Despite their presence in remote highlands, the wild yaks' domain has changed. Though people are few in those areas, mass carnage has reduced *drong* populations. Habitat use by the yaks now differs. Unlike a century ago, females today are more frequently found on steep slopes. Whether it's because of an individual's memory of rifles (unlikely), the safety afforded by extra rugged terrain (more likely), the removal of flatlander yaks, a culture of learning, or a reduced population density is not crystal clear. The possibility of a legacy effect, however, looms large. Balancing a dread of humans with the need for high-quality nutrition, coupled with providing safety for their young, are drivers of female biology. Yet, while wild yak cows rarely frequent alpine valleys and desert steppes now, males still do. Males' penchant for these ranges parallels that of the early era, when explorers first visited and persecution was low.

There are other sex differences as well, although I don't know if they've changed over time. Our surveys revealed that females remained, on average, about thirteen miles from the edge of glaciers, and males were twenty miles out. Despite its high elevation, the Tibetan-Qinghai Plateau is arid. Both sexes are averse to deep snow, but it tends to be light and patchy on the plateau and sublimates quickly, even in midwinter. In fact, so rare and so dispersed was snow during our surveys that the satellite pixels used in our analyses failed to detect it.

Female yaks were fully aware of the snow, however, and regularly nestled closer to snow patches than did males. When we measured actual distance, females were nearly twenty times more likely to be within two hundred yards of snow than were males. The question is why.

Might females be more sensitive to heat and, consequently, enjoy resting near or in snow? Perhaps, but our work was in winter, not summer, so that possibility is unlikely. What about body bulk? Males, after all, are at least one-third larger, and since size dictates metabolic requirements, maybe males just roam farther to access high-quality food. While the postulate makes sense, it begs the question of female proximity to snow.

If the riddle has an answer, it must be access to liquid. Water availability comes in many forms, but winter on the plateau is so cold that all but the highly saline lakes freeze solid. The valued liquid is quite likely to govern animal distribution, just as water does in hot deserts. Natives on every continent, human and otherwise, know this. As Przhevalsky noted in 1876, "wild yaks require plenty of water . . . when water is unobtainable they slake their thirst with snow." And we then worked out why females, even in winter, needed it more.

Our focus on the sexes allowed this additional insight, one we'd have missed had we just relied on satellite imagery and not included ground observations. Some females, we had noted, had distended udders. Despite ambient temperatures plunging lower than –20°F, mothers were lactating. Even in December, calves nursed.

Milk production requires water—accessible water. By situating themselves near snow, females converted snow to water, which ultimately meant milk for their calves.

Investigating natural history has bright moments, though perhaps not instantaneously. On the plateau, my senses had been dull—maybe it was the reduced oxygen at high elevation. Even though I had witnessed midwinter suckling, I had merely noted what I saw, without having drawn any conclusions about it. This was neither the way I was taught to do science (by hypothesis testing) nor the way I've instructed my students. But often, the reality of what we do is to simply record what we see. It is one way in which we come to understand animals. Only later—when talking with Ellen Cheng back in the United States, exploring our data, and rethinking the importance of observation as Darwin himself stressed—would I realize why wild yak females were situated near snow.

As important as the work away from the field can be, seeing the field through the eyes of the animals themselves will never lose its relevance. George Schaller had formerly admonished me for accepting my chair position at the University of Montana, where I'd teach classes in science and conservation. He felt I'd compromise my ability to do fieldwork, though not my love for it. Nature fascinates and resonates with me, though it doesn't for everyone.

I once asked Theresa Laverty during her PhD defense at Colorado State University how she might explain to a cynic of science how the natural world affects society. She answered seamlessly, saying that air and water are needed for plants to sustain wildlife, and they subsequently allowed the first humans to endure as well. Forests and grasslands, and animals big and small, are all interdependent. Human hunters had to use science in their pursuit of quarry. Success was not just random. Agricultural expertise followed, and across the millennia, our understanding of genetics and medicine also grew. All of this knowledge initially derived from the observation of species in nature and correlating their behaviors, which transformed into deductive reasoning, and finally experiments.

Nature has been critical in fomenting science. But we've gotten lost along the way, forgetting about curiosity and pure love of discovery that should be driving our researches. In his chef d'oeuvre, *Tracks and Shadows*, Harry Greene reminds us that it's not too late to rediscover our connections to nature, that we must stop and smell the flowers, despite the many pressing issues in the world waiting for resolution. We can do better. Understanding natural history is a first step.

The Tibetan Plateau is losing its capacity to store water because the Third Pole is warming. The Himalayas, Asia's water towers, have been a magnet for water, but a regime change is well underway. As snow grows increasingly patchy, wildlife, including female yaks, will have to find new winter water sources. As glaciers recede and runoff patterns change, yaks will go higher. Initially, this will appear to be positive because increased melting creates alpine meadows. But with continued deglaciation, preferred habitats will degrade and fade away through desiccation, and the permafrost will fracture. At some point, the yaks will not go higher because there will be no suitable habitat remaining. For muskoxen it's similar, but rather than altitude it's latitude. When they run out of room, their options end.

The biologies of yaks and muskoxen are steeped in refrigeration, and we know little of these species' resilience. If we wish to carry out conservation, we will need to know more. But there is also more to conservation than understanding pure biology. Yaks need ample space, protection, and suitable habitat. Conservation of these species requires more from humans than just protecting animals from poachers. There is planning about boundaries, livestock, and disturbances. Though the humans on the plateau are not densely concentrated, their livelihoods must be taken into consideration as the grasslands provide their sustenance.

Most of the world's population, in fact, lives near coastlines, not high elevations. Nearly two-thirds of the planet's 7.5 billion people concentrate within 250 miles of major water bodies, about one-third reside below 350 feet in elevation, and more than 75 percent of the global population lives at less than a half mile above sea level; few are three miles up. Humans are not increasing at high elevations at the devastating rates seen elsewhere, yet the challenges to them on the plateau remain serious. Water retention, cracking permafrost, feeding the family, and earning money are just some of their concerns. Other than earning money, the survival of wildlife species involves the exact same concerns as those of humans—food, safety, mates, refugia from inclement weather or a burning sun. But, without improved conservation for wild yaks, the next generation will know them only as lifeless etchings on rock.

Drugay, or DJ for short, is an ethnic Tibetan born on the plateau. He's also a budding ecologist who strives to conserve mammals. DJ uses his Han name Zhou Jiake on his passport. It's easier for authorities and simplifies his life during travel outside China, something he did for three years. I, however, know Zhou as Drugay. He was an undergrad at the University of Montana and was befriended by Ellen and Lishu when all of them were students there.

In Xining, the four of us reconnect at an internet café after our fieldwork at Kekexili. I cut pastries with my Swiss Army knife. As it turns out, DJ has been working for the World Wildlife Fund (WWF) on grassland issues but he's now interested in graduate school and will apply for the George B. Schaller Fellowship at the University of Alaska. A year earlier, he met Schaller in Montana. My letter of recommendation offers this characterization:

> He knows the bite of cold. He knows Western foods. He excelled while living in Montana. I never saw Zhou sad or in culture shock. He rides bicycles in cold and has cross-country skied in the northern Rocky Mountains. Zhou is tireless in work ethos. He has studied and co-authored a paper based on data he helped gather on Tibetan foxes. Zhou has worked with top-of-the line NSF-funded American biologists (Richard Harris and Andrew Smith) in western China. He has watched kiangs and

chiru roam across the Tibetan-Qinghai Plateau. Zhou's language skills are splendid and we have interacted and laughed.

Having just finished this field season, I feel better about prospects for wildlife in China, at least in western regions. Massive protected areas exist on a scale duplicated nowhere on the planet except Greenland. The police I've worked with are committed. DJ touts the younger generation's interest. The old guard is willing to talk conservation, as are newly appointed officials. Lishu won an Intel Corporation Award on behalf of WCS for her app on wildlife trade that helps Chinese police and military decode illegal animal products. DJ was awarded the Schaller Fellowship, moved to Alaska for his PhD, and will return to his homeland. Chinese postdocs study snow leopards and map tiger locations.

At the airport in Beijing, I check my luggage. The officer in charge of screening discovers my Swiss Army knife and informs me it will be confiscated even though it's in stowed baggage. My rationale fails to win the dispute.

I call for a Korean Airline official, hoping that my loyalty as a passenger will count for something. He concedes and explains my logic to the Chinese official—if the knife is in secure checked luggage it can't be used. That, too, fails. Then again, I have no idea what was really said as I speak no Mandarin. On my side, however, is time—hours of it until my next flight. Three more officials arrive and then two in plain clothes. Finally, they tell me my luck has run out.

I have one chance remaining. Now that two Korean Air officials are present and both speak English, I ask the second for assistance. "Please call the Beijing police and ask them to arrest the X-ray official for theft." To be certain my request is understood, he asks me to repeat it. I do just that, asking for the police to arrest the knife thief.

I'm pretty sure he's followed my instructions, because the gathering becomes highly animated. Six officials talk really fast. Two are getting mad. Hands gesticulate, and eyes glare, mostly at me.

A policeman in a different uniform arrives. The Korean Airline spokesman is calm. "Sir, they say if you want to go through customs and catch your flight you must leave now." Like a puppy dog with a tail between legs, I shuffle off begrudgingly.

Twenty-four hours later, I'm at the Missoula Airport. I collect my duffels. There is a gift inside, one from China—my bedeviled knife.

PART III

Gobi Ghosts, Himalayan Shadows

Geography has much to do with diversity and extravagance. It combines accidents and history.

Here's a quiz, probably best for your friends. What single country has moose and elephants, tigers, and caribou? If unsure, narrow things down by adding brown bears and leopards, and then wild camels. Befuddled? Include truly well-known icons like the graceful spotted giant flying squirrel and humped-back dolphin, and the bulbous-nosed takin or proboscis-swinging saiga. Now add the pygmy slow loris. An answer will assuredly emerge once we cap the list with giant panda.

For many species, a double whammy is temperature extremes in either direction. Consider, for example, us *sapiens*. Eskimos have the cold. Kalahari's San hunters have the heat. Yet we're all an interbreeding human family. Our nonhuman relatives are no different.

Ibex cross the arc from Africa and Asia to Europe. Populations of that species dot parts of Egypt and Ethiopia and range from Russia to Switzerland. In the former two, ibex are scorched, while in the latter countries it's cold, with temperatures varying from more than 110° to −40°F. Within one ibex population—in Mongolia, for instance—individuals can experience the same hot and cold extremes as those from the distal edges of their entire distribution. The same is true for Bactrian camels and numerous other species.

Need we seriously worry about a species' ability to withstand high temperature variance as the world warms? Might resource restrictions, like the increasingly limited availability of snow or water, be more worthy of concern? Other issues call out for our attention and study as well.

Counting individuals of a species, pedestrian as it may appear, is one way to start answering these questions and approaching these issues. How many individuals of species *X* or population *Y* are needed to assure viability, or, how few can remain before we worry? What do the population trends foretell?

For relatively well-studied species, the answers to these questions are generally known. Aerial censuses for elk in Wyoming or African elephants offer clues about their total numbers. Counting, however, is far from routine in the deserts and mountains of Central Asia. Government planes are unavailable, and private aircraft are prohibited. On the Himalayan slopes of Nepal and Sikkim, Bhutan and Arunachal Pradesh, forests obscure visibility. It's tough to know if species need conservation when little is certain about population trends or sizes. Yet, it's across these very environs—hot and cold—that an array of stunning mammals exists.

Sambar deer, spotted deer, and serows frequent forest glades. With these come tigers and leopards. Beyond the Himalayas are wild two-humped camels. Farther afield, the world's most northern antelope, saiga, roam concealed in valleys isolated by Mongolia's mighty Altai. Golden hordes—not the horsemen of Chinggis Khan but the white-tailed (also called Mongolian) gazelle—dwell to the east in the world's largest unfenced grasslands. But ascertaining the number of any of these animals is a colossal impediment to understanding a species' status. Without such figures, as well as a knowledge of threats to their populations, pragmatic conservation agendas will languish.

Some pathways forward are obvious—studies, habitat protection, and law enforcement. But, what if a species' status has remained unknown? Conundrums come in different forms. What if money is accessible for something sexy—say, captive propagation, the practice for pandas—but not for something conventional, like protection or counting?

Do we take the money? What counts?

The Kirghiz steppe trembles under the hoofs of countless Asian asses and herds of saiga... these animals move in such large herds that they cover the whole field of human vision.

M. ZHITKOV, "A Trip to the Kirghiz Steppe," 1849

10

Counting for Conservation

Explorers often capture the thrill of adventure and unusual observations. Population estimates differ in that they carry little of the fanfare that accompany such journeys. Still, decades before Zhitkov's journey between Siberia and the Himalayas, two Americans crossed uncharted prairies on another continent. Neither Lewis and Clark nor anyone else could estimate bison numbers. Historical guesses of the hemisphere's largest land mammal were ten to fifty million. Typically, explorers were simply fascinated by the animal itself, regardless of their numbers or the continent they inhabited.

In 1922, that fascination was coupled with the chase. Roy Chapman Andrews, leader of the Central Asiatic Expedition, writing about khulans, said: "It was our first sight of a new animal and one we had come far to seek. While we were still half a mile away, they began to run west by south going

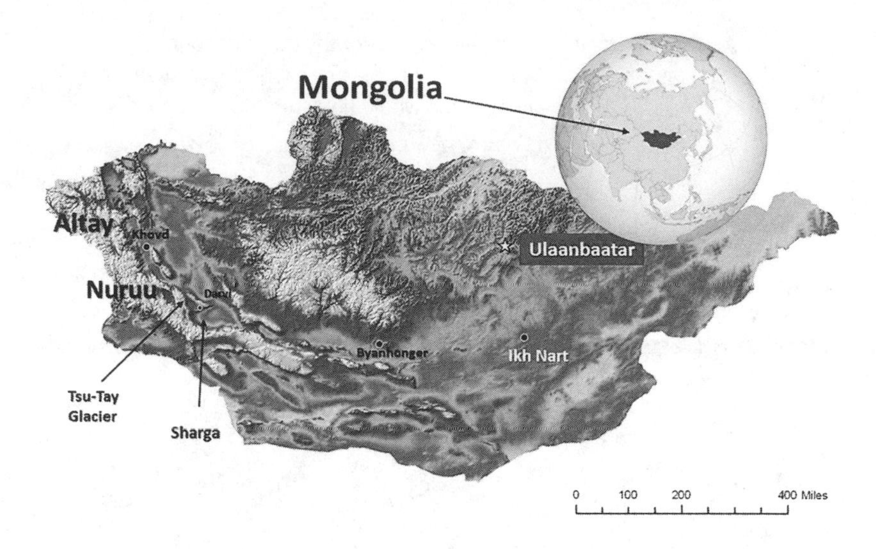

rather slowly.... Suddenly they disappeared in a shallow draw." Andrews was in Mongolia's Baga Bogdo Nuruu to bag specimens.

Some seventy-seven years later, I'm in the nearby Altai Mountains, but to count, not to kill. Sometimes counting and killing become entangled.

I'd later learn how much.

Outside the door of my yellow Bibler tent, the land seems liberated, vast and open. Far off, dust trails a motorcycle. A truck follows. More dust. Both vanish below a ridge.

I'm reminded of a prior trip, a different Mongolia late in the 1990s. It was clear and cold, and an October sky was growing dim. A lone star shone. Snow covered the ground. I heard hooves pound and a single rider appeared on a small horse. Our eyes locked. His song pierced the lonely steppe. Then the nomad disappeared. In eighteen hundred miles I traveled, there were no fences. Motorcycles were few, trucks fewer.

Five years have passed since my brief encounter with the horseman. Next to my tent today is a *ger* (Russian for yurt), which is used for cooking and sleeping, and it shields my Mongolian coworkers from Siberia's angry wind. To the north, autumn's first snows blanket the rising land. The Sutay Glacier stands sentry on the flanks of the Altai Range, its dome glowing in the shafts of a setting sun. Ibex roam its jagged canyons below us, never quite secure from a silent cat that stalks above. At one time, lower down on the steppes and in deserts would have been Przewalski's horses, wild Bactrian camels, khulans, and saigas. Only saigas remain there today.

In this land-locked country between Russia and China, my goal is to count and to capture the proboscis-dangling antelope. The Mongolians call it *bukhun*. It appears to be an impossible cross between an elephant shrew and a moose, but with the dangling proboscis of a tapir. The rest of the world calls this animal a saiga, and its babies are unquestionably more adorable than Bambi. Once spanning three continents, their preferred habitat was arid. It still is. Fossil remains are found from Canada to England. They also come from the nonglaciated islands of the East Siberian Sea and the Beaufort Sea above the Arctic Circle; saigas were once diminutive antelopes alongside massive mammoths. Far to the south, the fossils come from the Caucuses and France. Today the saigas' dwindling homes are the deserts and arid steppes of central Asia. Until recently, China had them in the wild. Five countries still do—Turkmenistan, Kazakhstan, Uzbekistan, Russia, and Mongolia.

The geographically isolated saigas of Mongolia differ genetically, resulting in size and dentition disparities from those farther west. They're a distinct subspecies. Some biologists, including the Gobi's most famous, Dr. Sandvin Dulamtseren, believe they're a separate species. Regardless, both forms have dropped precipitously in population size. I'm keen to study the long-snouted, bulbous speedsters, and no one knows how many persist.

Namshir, a veterinarian, is at our base camp. Of Buryat descent and born in the mountains near the Russian border, Namshir charms us with sturdy Buddhist beliefs. We'd met five years earlier when surveying wildlife in the Gobi and then again in the Tetons of Wyoming, where we crossed swollen rivers to radio-track moose. With us also is Buuveibaatar, a recent graduate of the National University of Mongolia, who is quiet, strong, and keenly interested in conservation. We lack knowledge about how livestock and people affect saigas. We know little of saiga survival and migration. Namshir and Buuveibaatar will bolster our team's effectiveness.

Severe wintry weather seals thc landscape, turning life into a deep freeze. Known as *zud*, these extreme wintry events kill millions of domestic stock in a single dose as the cold strips away calories from the animals and snow restricts access to their food. Effects on wildlife are less clear. If saigas escape *zuds*, where do they go and how do they get there?

Prior to the dissolution of the USSR late in 1991, nearly one million saigas roamed Asia's central steppes. Ten years later, 95 percent of them were gone, their horns marketed to China for medicinal purposes. But that can't be the whole story since only males have horns. Females, therefore, have little therapeutic value to the Chinese. Maybe they were eaten? Actually, the male haul was so great, reported British biologist E. J. Milner-Gulland, that females were no longer likely to become pregnant.

In the 1980s, when saiga populations were at their zenith in the USSR, photos of the hacked faces of Africa's rhinos were bleeding into the global media. The World Wildlife Fund floated a proposition to diminish the massacre by substituting saiga horn for rhino horn in a systematic and sustainable way, with the goal of satiating the Chinese thirst for curatives. This proposed use was for a species that, by the way, would later, in 1987, became protected under Chinese law. That protected status didn't matter, though: saigas were slaughtered anyway, as rhinos continue to be.

At the outset of the twenty-first century, a new suggestion came forth. This one was from the Mongolia office of the World Wildlife Fund in Ulaanbaatar (UB). The aim was to rescue the species not only by protecting it from poaching but also by enacting captive propagation. The idea was worthy of serious consideration. At international workshops in 2004 in the city of Almaty in Kazakhstan and in UB, one summary statement read: “Given that [Mongolian] saiga extinction is an immediate threat, it is necessary to prepare a carefully planned captive breeding and reintroduction program.” Delegates from ten countries attended.

In Almaty, only Bill McShea from the Smithsonian and Amanda Fine, WCS’s Mongolian director, questioned the need for captive propagation. They focused on numbers and the claim from the workshop that there were “less than 800 Mongolia saigas left.” If the dainty antelope could not be counted with some certainty, wasn’t it possible that the million dollars needed to finance a center could prove wasteful? Then again, if numbers were in fact precariously low, perhaps the need was justified. It wouldn’t be until 2006 that I’d try census techniques on saigas that were widely practiced elsewhere in the world. I was also intent to radio-collar them.

There would be a problem, however, and a big one. Adult saigas have never been captured or collared, not in Mongolia and not in the wilds of Kazakhstan and Uzbekistan, Turkmenistan, or Kalmykia in Russia. That specter will turn to reality when we confronted the difficulty of collaring.

In Mongolia, the night grows cool. Raindrops pelt the tent. In the *ger*, a slaughtered sheep awaits the appetites of our research crew, its entrails dangling from wooden lattice. I’m uneasy about our impending operation, because I’m unable to forecast prospects for success and uncertain if we’ll inadvertently kill a saiga. In lieu of a savory sheep-based meal, I try sleep.

Mongolia’s grasslands once teemed with wildlife. Marmots, pikas, and ground squirrels were thick. So were steppe lemmings, sandy-haired gerbils, and, arguably the cutest of all, long-tailed jerboas with their Mickey Mouse ears and adorable whiskered faces. At least forty different kinds of raptors were attracted to this furry buffet. Among them were steppe and golden eagles and buzzards. There were peregrine and saker falcons. Lammergeyers floated overhead, and black vultures and griffon feasted on carrion. Humans made use of the larder too; they trapped twelve-pound marmots, fire-roasted them much as American Indians had cooked porcupines, and then sold the hides. Herdsmen on horses, and sometimes camels, sheared sheep and goats and milked yaks and camels. A treat for them was *airag*, a gin-and-tonic equivalent made from fermented mare's milk.

Mongolia's transition from Soviet-style socialism to free-market economies in 1991 meant lean times as capitalism took hold. When I had first arrived at Sükhbaatar Square in UB during the 1990s, the ambiance was cold. Steely block buildings reflected Russian architecture, and statues of Mongolian leaders in military garb celebrated victories. Restaurants were scarce, their few patrons reminiscent of characters from Star Wars—thickly dressed denizens from the outback, warriors from Poland and Germany, and people ready to brawl. Winter nomads from the *ger* district made their way to the state department store in traditional clothing—furred hats, caftan-like *deels* draped to the ground, and leather boots. Coal-fired power plants belched smoke, turning skies gray and acrid. As the coldest capital in the world, more frigid than Stockholm or Moscow, few complained about the trade-off between heat and air quality. But the particulate matter coupled with a burgeoning car population still makes UB as polluted as Beijing.

Mongolians knew how to navigate the tricky streets. Foreign visitors did not. Manhole covers mysteriously disappeared, leaving gaping holes for the unwary. The night also held danger, as subterranean refugees who eked a life in tunnels would suddenly materialize, dangerous and ready for action. After two of my colleagues from WCS were beaten and one hospitalized, I smuggled in pepper spray to carry. A female coworker used capsicum, nailing an attacker squarely in the face.

Despite the tough social scene, wildlife remained abundant yet at that time. Red deer grazed along the Tuul River, which ran through UB. Just south, wild boars lived in forests. It was during this period that efforts to understand Mongolia's abundant wildlife began to take shape, by both researchers within Mongolia and those beyond.

Among the few Americans who knew the wildlife there were the tireless Schaller and the equally untiring Peter Zahler from WCS. Kirk Olson, a PhD student from Massachusetts also studying Mongolian wildlife, was focused on white-tailed gazelles—reddish-yellow beauties that spread from northern China into southern Russia.

Yet another American began work there in 1994: Rich Reading—fit, savvy, and fresh off a PhD from Yale. Soon thereafter, he invited me to help with a khulan project.

For over twenty years, Rich developed relationships with local universities and government and hired local biologists, including Ganchimeg Jamiyansuren Wingard, who goes by the name of Gana. They conspired, built an all-Mongolian team, and focused on an area called Ikh Nart. In 1994, it was a reserve on paper only and lacked real protection in the form of rangers; the desert enclave boasted half a dozen reptiles, three dozen mammals, and more than a hundred species of birds. For most, there had been little study and their statuses were unknown.

Today, the situation differs. Upgraded to a nature reserve of national standing, the United Nations Development Program touts Ikh Nart as a model of success. Among Rich's unsung awards is one from the government as Mongolia's best foreign conservationist.

Rich and the team of Mongolian researchers, including Gana and Amgalanbaatar (known as Amgaa), were able to bring local herders on board to value protection, habituate argalis and ibex, and expand reserve boundaries. More than thirty Mongolian graduate students were trained, and individual studies were conducted of pit vipers, cinereous vultures, hedgehogs, bats, wolves, and foxes. A women's collective market was developed to sell crafts to tourists. These activities created economic benefits for local residents near the protected area. A conservation education center, facilitated by Rich, continues to promote outreach.

Following my first visit to Mongolia, I became interested in developing a collaborative project. But when I asked what I might contribute, government and nongovernment organization (NGO) leaders suggested I look Elsewhere: Ikh Nart was covered. So were khulans, thanks to Austrian ecologist Petra Kaczensky and a talented wildlife veterinarian, Chris Walzer. Mongolian teams studied wild Bactrian camels. Przewalski's horses had been reintroduced to the Hustai Nuruu, and the tiny population of Gobi brown bears was already under scrutiny.

In the low deserts to the west, however, was a species ignored. This was the *bukhun*, or saiga.

I anxiously sought the advice of Dulamtseren, the dean of Mongolia's mammals. Now retired, the professor was kind, though formal with the demeanor of a Russian academic, and expressed a palpable love of *bukhun*. We'd been in the field together years earlier and even coauthored a paper along with Gana and Rich on mass ungulate mortalities in the Gobi. Upon seeing me, the professor's eyes welled up: "For four decades I had hoped we could learn more about saiga. Now we have that chance. I am so happy you are here." He then introduced me to another Mongolian Academy of Sciences member, a different Dr. Amgalan. His colleagues joke his snout is like the species he studies, the saiga. He's not to be confused with the other Amgalan, the argali expert who we called Amgaa.

To find saigas, it's a three-day drive from UB across rutted tracks and virtually no signs to the Govi-Altai Province, which is similar in size to Nepal. These remote desert and mountain lands include few humans—less than one person per square mile, making this a good area to investigate saigas in the wild. North is the town of Khovd. It's located in the province of Khovd, where a 1925 photo shows saigas migrating through the languorous capital. To the east is Khar Us Nuur National Park, a vast wetland complex with steppe but few saigas. Our target instead will be the Sharga Nature Reserve, another so-called paper park at the foot of the Altai Mountains. ("Paper parks" are protected areas that exist on maps and in legislation but offers little real protection.)

Enkhee, the owner of an old, beat-up, four-wheel-drive Russian van knows the way. From UB, we cross grassy steppe where horses roam freely and *gers* are few. We angle toward Bayankhongor, a sleepy government hub with a languid downtown square where we will spend the night en route to our destination. Our first priority there is to lock the van out of sight to prevent local theft. Later, at the only open restaurant, several drunks try to assail a female coworker. One of the drunks is slammed into a wall. We continue the next morning toward the Sharga Nature Reserve and watch helplessly as someone else's jeep gets stuck crossing a bloated river. Enkhee is on a mission and keeps driving.

Rugged mountains materialize. The valleys here contain black-tailed gazelles and escarpments for ibex. We hike and watch two wolves chase the horned wild goats otherwise known as ibex. By day's end we've netted seven skulls: three argalis, two ibex, one marmot—and one human. Like Enkhee, we keep going.

The land grows harsh. Enkhee has a knack for navigating rock-strewn terrain and remains oblivious of that fact that we're being jolted from seats in his seat-beltless wonder. Following a few more hours of bouncing, my head now bloodied by the ceiling, a world emerges that is more reminiscent of the Chinggis Khan era than today. A bucolic profusion of camels, horses, and *gers* appears on the horizon. Greetings are exchanged between us and the local people. A woman approaches with tears streaming down her cheeks: she's not seen her son Enkhee for five years.

Enkhee's mother brings snacks, including generous chunks of raw fat and fermented mare's milk. I ask about our hosts' lives out here and am not surprised that World War II passed by them without notice: they learned of it only five years after it ended. In the morning, I learn to ride a camel.

A day later, we meet local people in Darvi, the *soum* (small town) closest to Sharga Nature Reserve. At only three thousand feet in elevation and in the rain shadow of the Altai Mountains, precipitation is two inches annually—half what desert-based Las Vegas receives. Yaks, cows, and horses wander unpaved streets. Solar panels generate electricity, but power is limited to a few hours daily. We find the *soum*'s (or town's) spigot. The pump is not turned on and no one knows the whereabouts of the man with the key. We wait hours to fill our water jugs. Young girls in school uniforms pass, stop, and stare. Two kind drunks help us find the mayor, whose cooperation we will need for the saiga project.

At about the size of Alaska, Mongolia has a population four times larger, yet outside of Ulaanbaatar, it remains among the least-inhabited countries per square mile. Most Mongolians live in the city. Nowadays, much land not only remains empty of people but is also emptied of most wildlife.

The profligate growth of coal, copper, and gold mining since the 1990s has meant more ninjas—the local name for miners living frugally on the land and taking what they need—and increased poaching. There are more roads, and by 2005 fencing had become progressively evident. Beasts fueled not by grass but by fossil fuels replace camels and horses. Wildlife once abundant in most valleys en route to where we'd see saigas in the Shargyn Gobi is now rarely encountered from UB to the Khuisiin depression, just to the north of the paper reserve Shargyn Gobi. In their 2006 censorious review, "Silent Steppe: The Illegal Wildlife Trade Crisis in Mongolia," attorney Jim Wingard and biologist Peter Zahler report red deer numbers are down some 90 percent since 1995; the same is true for argalis, and likely for boars. Little is known about moose or musk deer, but the slaughter of marmots was so intense the animals have been offered solace by a three-year federal trapping moratorium.

Such losses, though, bring opportunity. Unlike governments in other parts of central Asia, Mongolia has little angst about foreign NGOs or radio collars and applauds exposure of their scientists and students to the outside. This bodes well for WCS and our pursuit of saigas.

Saigas are quite small, about as tall as one's hip. At long distances, they fade from sight. I'm talking with Amgalan, the Academy of Sciences top saiga researcher when he spots them at two to three miles, though I cannot see them. I've never sighted animals before he has. Fortunately, I'm with Julie Young, a talented WCS postdoc with eyes almost as good as his.

Amgalan explains the simple algorithm used to estimate population sizes. Every animal is counted. The key assumption is that none are missed, and then an adjustment is made for the size of the sampled area relative to a total region. The calculation results in an "absolute" number with no variance. For many population biologists, this method is a problem because the uncertainty of the estimate is not recognized.

I discuss my concerns about this with Amanda Fine and others from WCS. If an animal is close to a transect line (a line that cuts through a habitat and along which a researcher moves), its probability of being counted is far higher than one that is, say, two or three miles away. Detection is inevitably imperfect. The counting techniques used in Mongolia might be accurate, but knowing the margin of error would help to give a sense of how reliable such counts are.

What if many saigas remain uncounted? *Would that obviate the need for a captive center?* This begs the question of whether money that could be used for the facility might otherwise be plowed back into conservation. One approach has zest to funders—the center, as that would seem like real action, the act of doing. The other approach, counting, is far more mundane and it might also involve hiring rangers. Like everyone else involved, my main concern is determining the best way to sustain saigas. If habitat exists and numbers in the wild are decent, why not protect them there?

Scientists are trained to ask questions, even if doing so upsets some people: if they aren't asking questions, they aren't doing their jobs. The ex-Soviet system, however, stifled inquiry, so that although Mongolian biologists were well trained to ascertain facts, questioning authority was not always the best path.

Chimed-Ochir and Onon from WWF are each strong proponents of the captive propagation maneuver. When I ask about errors in sampling, I receive harsh stares in reply. Chimed tells me he spots saigas at four miles. His English is perfect, so, for fun, I ask if he means kilometers. He pauses for emphasis, and grins, saying "miles," as if to imply I'm not up to the task.

In thinking how best to do saiga surveys, we immediately rule out aerial transects. Although volunteer pilots from the United States and Germany were anxious to help, getting a small aircraft into Mongolia would be a nightmare. Rich Reading had attempted it, but that was years earlier in a behemoth aerial tanker.

We opt instead for a method called distance sampling, which is based on probability theory. The first assumption is that the probability of animals along a transect line being detected is 100 percent. Just imagine an animal in front of the vehicle that is driving down the transect line. You hit it, bingo. One animal has therefore been counted. Then, the counting becomes more complicated, hence the second assumption—the farther the perpendicular distance between the line and the animal, the less the probability is of discovering it. Transects must be randomly distributed across the landscape not merely to establish easy driving but also to reduce the introduction of bias. The methods have to be accurate, repeatable, and offer a measure of precision.

With a field crew that includes Buuvei (Buuveibaatar—the recent grad of the National University of Mongolia mentioned earlier), Julie, and Namshir (the veterinarian), we establish twenty-four transects in the western Gobi during early fall in 2006 and 2007. Three miles apart and nine miles long, some transects are inside the paper reserve and others outside. Where topography is tortuous and tires pop, we complete observations for only a single transect in a day. Some days we manage three. The surveys take us into mountains, through sand dunes, and across rock. Where there's water, there's mud and we get stuck. Rumors of Americans being in the vicinity spreads via the nomad pipeline, reaching the desert towns of Tonkhil, Togrog, and Darvi. We explain at each why we count saigas.

We shift base camps frequently, sidestepping aggressive dogs, occasionally in packs as large nine that are chasing saigas. Friendly herders with flocks of up to a thousand goats and sheep, as well as dirt-poor herders with fewer than three hundred, invite us to their *gers* for meals. Nearby, griffons and cinereous vultures feast on dead goats. At camp, I introduce Frisbees to the WCS staff and soon they are near experts. At night, the group stays up, play cards, and tells stories. The camps are deliberately alcohol free, though sometimes the evening camaraderie is shattered when wayward herders arrive drunk. More often than not, they don't tarry but go on their way, in search of more vodka.

One early morning, something hunkered down on a stony hillside draws my attention. Perfectly camouflaged by its lavish fur and a darkly ringed bushy tail, the Manul—sometimes known as Pallas's cat—peers intensely in a defensive behavior common to small desert felines. One herder that we met had a fine jacket that must have been stitched from twenty not especially prudent of these cats. A few days later, a motorcycle bounds along a dirt spoor, ridden by an apparently skilled rifleman: he shoots a Corsac fox. Life is tough for foxes, cats, and herders alike.

On an off day, Buuvei and I elect to climb Sutay. From its thirteen thousand–foot summit we hope to glimpse the Tian Shan Mountains, knowing it's actually unlikely that we'll be able see two hundred miles southward into China. We wind our way up narrow gorges and deep canyons still in shadow. Ice clogs a quiet stream. We flush chukar partridges. The snow builds. Imprints of three sets of *irbus* (Mongolian for snow leopard) are imprinted in the snow—a snow leopard and two kits. There's a marmot, fat and evidently dead, its leg cut to the bone by a steel trap. *How long has the prisoner suffered?* I lean down for a better look. The marmot thrashes in a violent struggle, its leg dangling. Somehow we free it and it scampers away.

We find skulls—yak and argali. We see ibex, alive and abundant.

Whereas the sun warms the snow, evening cools it. The wind leaves it glazed and sculpted. The glaciated dome of Sutay rises before us, looking inviting. All we'd need is one last push to reach the summit. Crampons and climbing axes would help with that but we lack that kind of gear and since dying on this holy mountain is not my wish, we retreat. That's OK with Buuvei. He's more interested in the biodiversity below.

With transects complete, and analyses underway, we begin to make plans for the next few years. Our aim is to develop workshops and more conservation engagement.

We'll ask herders themselves about saiga population sizes. We'll also ask what the Ministry of Nature, Environment and Tourism would like to see as target populations. And, to expand further, we seek members from the Mongolian Academy of Sciences for suggestions about other invitees to a conservation workshop. Two names come up immediately—Chimed-Ochir, the man who can spot saigas at four miles, and his WWF sidekick Onon—but they are people already on our list. To round out attendees, we add the head of police for the Altai Province, local mayors, Mongolian's Environmental Protection unit, and the military because they deal with illicit trade.

The diverse attendees will also include rangers, military, and herders. Those from WCS will include Amanda Fine, Julie Young, a splendid translator named Narantsatsral, who later turns out to be equally adept at Frisbee, along with a couple of other Westerners. Given the seriousness of poaching, the governors of the Altai and Khovd Provinces—Tsogtbaatar and Naymdavaa, respectively—will join us, as some of Mongolia's top brass. The secretary of the Convention on International Trade of Endangered Species will also be there.

While the National Geographic Society and WCS had supported the initial research, we approach the Trust for Mutual Understanding, a New York City–based nonprofit that backs cultural and environmental exchanges. They approve workshop funding. We select a past-its-prime Russian resort up in the mountains near Altai, a two-day drive from UB, but the location makes it possible for the herders and local officials to attend.

Set in a larch forest high above the desert floor, the setting proves ideal. There's a cavernous room to serve as a cooking and meeting hall, unheated bungalows, wooden beds, and a loud generator for electricity. Though neither running nor heated water is available, a small creek bisects the remote property. Domestic yaks wander freely, which means that, for those walking about at night, there is a clear and present danger: fresh yak patties.

The plan is for there to be a few talks given and for the entire thirty-person audience to then form subgroups and address key questions: Where should saigas proliferate? What are the factors that limit populations? How many saigas should populate the landscape? The ultimate goal is to develop consensus about paths forward.

With a welcome by Amanda, self-introductions, and twin addresses by the governors of the Altai and Khovd Provinces, the morning has evaporated. We break for tea and wander among yaks in the energizing mountain air. Following the break, presentations covering local viewpoints, economics, livelihoods, and saiga biology are given. Saigas have a remarkable capacity to reproduce. Females can conceive at six months of age, and adults in good condition can actually produce triplets.

The talk on population estimation explains both why we're here and why we need rigorous sampling. We carefully avoid mention of the frailties of methodologies used previously and praise the efforts of those saiga researchers that preceded us. Then we expose the meat of the problem.

The population estimate for saigas in 2006 was 4,900 animals and, in 2007, 7,200. We emphasize that any approximation is only as good as its margin of confidence. There is a 95 percent probability that the number of saigas is between 2,762–8,828 (2006) and 4,380–11,903 (2007). These estimates apply only to the Sharga study region, which is about one-third of Mongolia's saiga range. While the variation is large, it's clear there are more saigas than anyone suspected.

Chimed-Ochir's hackles rise as he senses a threat to his dream of captive propagation. The presence of more saigas means there is less need for a center and, therefore, less need for money. His manhood has been called into question and so he delivers a timely assassination of our survey: "These are estimates, not complete counts. You cannot possibly be correct." *Obviously, he doesn't get the point.* "The findings do not hold value given such wide margins," he adds. To drive the point home, Chimed argues traditional Mongolian counting methods have been proven effective in contrast to something like distance sampling, which has been imported from the West.

I have to come up with a quick response. Should I say that Chimed is simply wrong or grandstanding? Given the tedium of translation from English to Mongolian and back, I'll kill the audience's interest if my retort is too long. So I ask why Mongolia encourages other scientists to help with population estimation and then pose a rhetorical question about whether Chimed accepts sampling error, yet considers assessments of Mongolia's human population size valid. Contemptuously, he suggests we pick this back up in UB when other academy members can weigh in. It's a good idea and will give us a chance to vet our findings broadly.

Over the next day other attendees are heard from, except for two: the governors have retreated to their common *ger*, where they drink, never to reappear at the conference. The consensus from the group about an acceptable population size is encouraging though, with a minimum target of fifteen thousand saigas. Poaching and *zuds* (severe winter storms) are identified as prominent threats to increasing the saiga population.

These concerns were similarly expressed back at the 2004 meetings in Kazakhstan and UB, although the participants then were not the same as those currently attending. A captive-breeding center was also recognized, at the 2004 meetings, as an excellent educational tool (if it didn't prove too expensive), but its broader conservation role was questioned. High numbers of sheep and goats were considered a serious problem for saigas, though no solutions were offered, other than to supply food for saigas during *zuds* and to maintain a few watering points free of livestock during summers. The establishment of safe zones for saiga movement was not something that had been considered previously. It was a topic we'd take up later. Corridor protection, as conservationists term it, was a well-known strategy.

Back in UB, Julie and I prep for our presentation to the Academy of Sciences, Chimed-Ochir and Onon of WWF, and others. The stakes are high: depending on how our presentation went, the result would be either that they would accept estimates of saiga numbers collected by methods they were unfamiliar with or that they would consign saigas to captivity for later reintroduction. It was far from a zero-sum game, however, given that there might be middle ground: perhaps adult saigas could be captured and collared without killing them; if this plan proved doable, it could facilitate better estimates about population sizes and movements. Unlike with muskox reintroduction, which historically involved adults being mercilessly shot in order to capture the young, saiga calves could be caught while adults remained unmolested. Young saiga calves, after suckling, simply drop back into the grass as their mothers move away. This would make for easy calf capture. The feasibility of capturing saiga young was clearly demonstrated by the Russians in the late 1950s, their having tagged a remarkable 18,279 saigas and releasing the babies back into the Kazakh wilds.

This approach was also tried in Mongolia when it was still under Soviet rule. In the 1980s, Dulamtseren and a crew captured more than a hundred newborns, rearing them in captivity with domestic goats. With the goal of reintroducing saigas to former habitats, fifty-four survivors were released. Sadly, the project was considered a failure "because of the presence of large amounts of predators such as eagle, condor (*vulture*), and lynx" in the area where reintroduction was being attempted. Saiga ranges did not expand.

Twenty-five years later, we ask what incentives exist for returning saigas to the wild once they're in captivity. Did it not make more sense to keep them free in protected habitat and to coordinate with sheep and goat herders to ensure the survival of wild saigas? Centers for holding animals captive certainly have extensive value, including education and science. The Center for Wild Animals of Kalmykia in Russia has eight hundred dedicated acres for work with saigas, including doing research, receiving local and international guests, teaching, and doing outreach. In the deserts of northern China, the Gansu Endangered Animals Research Center in Wuwei, where wildlife biologist Aili Kang had studied saigas, has bred Przewalski's horses and Bactrian camels, so in theory the same could be done for saigas.

But to argue that propagating Mongolian saigas at a hypothetical center is critical to their survival was a different matter. This is not a species whose wild population dropped to less than a few dozen animals, as had black-footed ferrets or California condors. Nor was the threat of extinction so imminent that the only option to rescue the species was capture, breeding, and rerelease. Our counting gambit was complicated—it was seriously compromised by local personality, cultural inertia, and politics.

At the acrimonious meeting in UB, Julie's and my central points had little cachet. A new tactic was needed. The issue was not just about saigas, it was how to improve in situ conservation. While we had made some inroads with workshops and meetings in Mongolia, our impact might be broader if we switched both approach and continent.

On behalf of WCS, I invite an array of ecologists from the Mongolia Academy of Sciences—the nascent and the established, young and old—for a training workshop at the University of Montana. We'll focus on population estimation and mathematical logic, engage with computer models, and stress practicality. Then, we'll go to Yellowstone and see how sampling matches reality.

In 2010, Amgalan (of saiga fame), Lhagva, head of the Mammal Laboratory at the Mongolian Academy of Sciences, and Buuvei board a plane for Montana. Gana, who had moved here eight years earlier, agrees to translate. Excited about their visit, the UB team is expectant of the service and civility for which America is often known.

Their flight schedules get off kilter, and the group reaches Missoula at 1 A.M. They had expected a fine meal on the Northwestern flight but had the pleasure of crackers instead. Famished, they ask for a real meal. They have arrived, however, in sleepy Montana, not stylish New York. Fortunately, I am able to find an open Denny's. We stop at a red light. There is no traffic, and the Mongolians urge me to drive through. I wait for the light to change, explaining that we don't ignore red lights in America. Moments later, we're at Denny's, and they have their first-class meal.

Over the next few days, we work long hours in my lab, use spreadsheets, and sharpen simulation skills. They digest the logic of distance sampling. Russian-trained and with plenty of math savvy, the biologists absorb the applied value of central tendencies, measures of variance, and probability. We head to Yellowstone. The first animal we see is a wolf. It's not running or frightened. It's not about to be shot. Smiles as big as the Wyoming sky sweep across their faces. We see bison, pronghorn, and elk. They note tourist-based economies, and watch visitors steeped in fascination with Yellowstone's geysers, scenery, and wildlife.

Throughout the next year, Julie and I receive notes of thanks from them. We're surprised to learn that the Academy of Sciences has endorsed such sampling schemes country-wide. The World Wildlife Fund has now invested heavily in protection and education. We were soon to work with their rangers, though the propagation center was never built: the counting we had done mattered.

> The methods you use are set up by the philosophy you have. If you think it's OK to let individual animals die, it can be a self-fulfilling prophecy. Wild animals are living on the edge. They can't tolerate stress.
>
> MARC BEKOFF, 2002

11 To Kill a Saiga

When Roy Chapman Andrews mounted expeditions to Mongolia in the 1920s, khulans "looked very neat and well-groomed in their short summer coats, and galloped easily." He went on to say that "the [khulan] stallion which we followed traveled twenty-nine miles before it gave up and lay down. The first sixteen miles were covered at an average speed of thirty miles an hour . . . [during which] there was never a breathing space." The brutally exhausted male was roped. Chapman didn't say how much longer it lived.

This kind of profound pursuit results in muscle necrosis, and sometimes even death, through a process called capture myopathy. Even the stress of simple handling can be injurious. Animal-care committees would never approve of the chase distance practiced by the likes of Andrews in the 1920s, and so my problem is how to capture saigas alive and not kill them in the process.

Unlike my virtual backyard in North America where pronghorn are herded by helicopters into nets, helicopters are not an option in Mongolia as they're banned there. Neither is ground darting possible: any unwary saigas were killed by poachers long ago. I can't look to anyone for advice on how to go about this as adult saigas have never been captured.

Kirk Olson, the PhD student also studying Mongolian wildlife, uses jeeps to drive white-tailed gazelles into nets. For ibex and argali captures, Mongolians can use horses as these wild dwellers of cliffs and slopes are not as swift as the dainty, fleeter saigas. Kirk is not using his nets and graciously offers them.

I wonder about animal safety: How far can I chase them without doing harm and how fast? I'm not anxious to learn whether saigas can out-endure khulans, which are some ten times larger than saigas. Smaller species overheat more quickly. No one has told me what happens if we kill a saiga.

The Wildlife Conservation Society secures permission for our research through the Mongolian ministry that governs parks. Onon of the WWF—who's never been thrilled with our project—informs her friends at the Ministry of Nature and Environment about our research. Their officers tell her to tell me that I'll be imprisoned if an animal dies. A Mongolian jail is a new thought.

I have nightmares of saigas that have been chased, legs broken, heat stressed. There's another bad dream that I have, one involving talons—vise grips tearing into my shoulder, blood exploding—the eagle's victory dance. The soaring serpent returns to the outstretched arm of a Kazakh hunter from the province of Bayan-Ölgii: revenge. For which there's only a single victim: me, the guy behind tortured saigas.

The capture plan involves deception. Stretch chest-high nets across the desert. Secure them against wind with upright sticks pounded into the ground. Keep the nets loose to enmesh saigas. Chase saigas for short distances only—and only in cool weather. Hope for no broken legs. Hope that not more than one or two get entangled at once, that the team does not behave like banshees during handling, and that the terrified animals are blindfolded during capture. Hope the collars are attached quickly and the animals swiftly released.

My other concern is for the safety of the team, something less on the mind of Mongolians. Vehicles here have no seat belts. Passengers will likely ricochet around the vehicle like microwave popcorn during high-speed chases. I hope drivers avoid boulders and each other: saigas tend to change directions like cats. I'll remind animal handlers to be delicate with saigas and not to tackle them as they would sheep and goats. My other hope is that animals will be near nets, not twenty-nine miles out.

Piece of cake. What could go wrong?

Two veterinarians accompany us. One is named Mike, from the U.S. Department of Agriculture. The other is Namshir. Most days Mike sits in his tent waiting for the action to begin. Namshir, in contrast, seizes every opportunity to keep busy. He is generally out looking for saigas, fixing collars, or cutting potatoes. We have three vehicles, including a high-wheeled Russian van, a durable Russian jeep called UAZ, and a Toyota Land Cruiser.

To entice saigas toward nets we import pronghorn from America. Not real, breathing ones—just life-sized mannequins. They'll be planted on the other side of the net from the saigas. When saigas see their white butts, we expect they'll race to the pronghorn, thinking they are brethren they can follow, and be intercepted by mesh. I'd have preferred a more realistic saiga model, but American designers wanted $950 to build a prototype. The pronghorn cost $75 apiece. A herd of five pronghorns therefor accompanies me to Asia.

Gobi means "big shallow basin," "desert," or "rocky" in Mongolian. It's a place of little rain, few humans, and extremes of hot or cold—a place similar to North America's Great Basin or Mojave deserts. The Gobi has saigas. Bactrian camels, wild asses, and Przewalski's horses.

Though their wild analogues once roamed the north in Beringia, on the cold, dry scapes of what is now northern Alaska and the Yukon late in the Holocene, the only spot to see these refugees now is the Gobi. Within the Gobi, argalis, ibex, gazelles, and snow leopards still wander the earth. The true deserts of North America lost their native Pleistocene camels and wild goats, as well as their native asses and horses. But bighorn sheep, the North American equivalent of argalis, are present yet in desert areas, as are cougars. Jaguars—big spotted cats of the Neotropics—sometimes hunt the desert highlands of southern Arizona, even when snow covers the ground. All of North America's deserts have a saiga equivalent: pronghorn.

Fleet and migratory, both species face threats more immediate than extremes of heat or cold. Although heavily poached and heavily hunted, saiga populations were up to maybe four hundred thousand by 2015. In May of that year, something tragic happened: more than two hundred ten thousand saigas died in a synchronous collapse—the most sudden decline of any large mammal known, anywhere. In less than a month, bodies were strewn for miles and miles across the steppes of Kazakhstan. A soil bacterium was suspected; its lethal effectiveness was likely prompted by unusual weather events. Mongolian saigas were unaffected then, but they've not escaped: early in 2017, they were hit by peste des petits ruminants—a highly fatal virus with plague-like effects probably transmitted by domestic goats and sheep—which took out about half their population. Also, when *zuds* sweep down from Siberia, Mongolian saigas need unrestricted access to food and areas free of deep snow in order to survive. And the documentation and protection of essential saiga migratory pathways is one of the reasons I'm back in the Gobi.

Like saigas, pronghorn were once drastically reduced, from millions to fewer than twenty thousand. To avoid a bison-like fate, restoration efforts began early in the twentieth century. Pronghorn now number some seven hundred thousand. But their future is not all roses: from Saskatchewan to Mexico, fences, roads, and suburbs block their seasonal migration pathways.

We've learned some tricks charting pronghorn movements. Maybe they can help saigas.

"You're not from here, are you?" asks the discerning ex-rancher from Wyoming, now an energy magnate.

Cowboy I am not. I don't own Wranglers. My boots are neither tobacco stained from spitting chew nor pointy toed. My pickup truck is not a gas-guzzling monster. My small Toyota carries the innocent letters WCS on its side, which the observant Sublette County sheriff must have noticed: Jon Beckmann and I have been pulled over.

It's true that the truck's license plate is covered by snow. But so are the plates of twenty other gargantuan pickups that also passed the sheriff. But unlike my truck with its conservation logo, their trucks are manlier, reflective of the burgeoning gas fields that fuel the local economy. Fracking is the new mantra in 2003, and profligate growth the game. More than fifty thousand wells are being built yearly in North America, and the increasingly unsightly expanse that is home to many of them known as Jonah Field might be a good place for Jon and I to accomplish some wildlife conservation. Though it's where one of the continent's largest assemblies of machines digs deep into Earth's belly, the Upper Green River Basin is also home to a hundred thousand wintering ungulates—including moose, elk, mule deer, and pronghorn—and where we expect animal movements to be substantially disturbed.

A few years earlier, Steve Cain and I wanted to know where pronghorn went once they left the Grand Teton National Park, where Steve is the senior park biologist. Elsewhere in the Greater Yellowstone Ecosystem, 75 percent of their migration routes had disappeared. Elk had lost 60 percent of theirs, and bison 100 percent of theirs. If pronghorn from the Tetons could not access less wintry grounds, they'd experience the same fate. Identifying corridors and keeping them connected to critical summer and winter habitats were, and remain, essential conservation ingredients.

Though the general public doesn't delve into the subtleties of migration the way biologists often do, they understand the general outline—that is, getting from point *A* to *B* matters. If gray whales cannot find openings in the ice, or if birds cannot fly south from the Arctic, they die. When the passes leading out of the Tetons fill with deep snow, pronghorn perish.

The world's land migration routes are under serious attack. Across southern Africa, most zebra, wildebeest, and springbok migrations were gone due to overharvest or as a result of land conversion by the 1980s. Fencing truncates others, even in parks the size of Kruger or Etosha. To understand what, if anything, might be doable to ameliorate this situation, more needs to be known about the extent of the challenges.

Steve and I wanted to discover where Teton pronghorn winter and the routes they use to depart and reenter the park. Obtaining that kind of information would allow a conversation about conservation threats beyond park boundaries. We placed GPS collars on female pronghorns in Teton and waited a year for their seasonal movements to play out. Our results seemed to show that they used only a single route, and wondered whether this could possibly be true. But we had, in fact, discovered the one route used by all pronghorn.

The corridor included mountain passes above eighty-five hundred feet, forests thick with lodgepole pine, and topographical bottlenecks squeezed by cliffs and rivers. The narrowest passage was just 120 yards wide, the length of a football field. If passages like these become blocked, the migration pathway would collapse and the Tetons would lose their pronghorn.

The narrow migration route through which these animals move has long been known. Native Americans ambushed pronghorn at Trapper's Point, a killing field, where bones have been found dating to some fifty-seven hundred years ago. It's the same path still used by Teton animals. The timing of movements remains the same, something we know because fetal bones at Trapper's Point archeological site are at the same stage of in utero growth as those in pregnant mothers now passing through in late spring. The pathway continues to be of current biological and cultural value.

The annual movements we documented with GPS are impressive for other reasons as well. One female traveled more than 420 miles, a distance paralleling that between Boston and Washington, DC. When animals leave the park, it's as if they blast out, covering ninety miles in three days. No other land mammals between Canada and Tierra del Fuego in South America migrate such distances. In contrast to caribou or wildebeest, whose migratory swaths might be twenty miles across, Teton pronghorn use an invariant strip less than a mile wide. Knowing this, I sought help to develop a conservation strategy that would protect this last vestige of a single, long, and symbolic route. The challenges to do so, however, would prove formidable.

To start with, I'm not local. Much of Wyoming is as averse to outsiders as to wolves. Fossil fuels are currency and king. Big rigs and man camps (i.e., temporary employee housing for oilfield workers) block pathways for pronghorn.

To develop support, I meet several times with Wyoming's governor, Dave Freudenthal. I point out that Wyoming had the world's first national park (Yellowstone), America's first national monument (Devil's Tower), and its first national forest (Shoshone). Shouldn't Wyoming have America's first nationally protected migration corridor? I go to Washington, DC, to lobby. Mary Gibson Scott, superintendent of Teton National Park, Kniffy Hamilton, supervisor of Bridger-Teton National Forest, and Barry Reiswig, head of the National Elk Refuge all sign a public statement to support pronghorn protection.

The Wildlife Conservation Society facilitates public and private meetings. In Sublette and Teton Counties, some of those meetings are contentious. Ranchers and outfitters, wolf lovers and conservation haters, local businesses, schools, industry, and county commissioners attend. So do the local chamber of commerce and representatives of Wyoming's game and fish and highway departments. The feds—representatives of the National Park Service, the U.S. Fish and Wildlife

Service, and the U.S. Forest Service—are there, as are Amy Vedder and Bill Weber, both WCS field-savvy practitioners who successfully developed economic and social strategies to conserve Rwanda's mountain gorillas. The prominence of a pronghorn mural at the Sublette County Library and the pronghorn trophy heads adorning the meat department at the local grocer prompt Amy to coin the moniker Path of the Pronghorn for the migration corridor.

In my presentations, I share information about the deep cultural meaning and ecological prominence of the migration. I ask whether the "speed goats," so named by Lewis and Clark, are sources of pride or mere menace. I ask if the long-distance speedsters are worthy of protection as an American icon of global migrants. I ask if protective measures will be a useless gesture to enhance the Teton's three to four hundred pronghorn when the rest of the state has four hundred thousand or more.

Pundits, at any rate, had claimed that any protection for Teton pronghorn is a mere gesture, an emblem, and not helpful for pronghorn elsewhere. So, I close my presentations silently, no words or labels, just photos: a 1950s black-and-white of Rosa Parks seated alone in a segregated Alabama bus; the 1972 Pulitzer Prize–winning shot of Phan Thi Kim Phus, the Vietnamese girl running in horror, naked, from a napalm attack; John Lennon seeking peace; three-year-old John Kennedy Jr. saluting the coffin of his assassinated father; and Jane Goodall kissing a chimpanzee. Just symbols.

Support for the project builds. Local papers write that economy and conservation go hand in hand. I coauthor an op-ed in the *New York Times* arguing that if we want to conserve "where the antelope play," we need protected corridors. The *Smithsonian* and *National Geographic* carry pieces on pronghorn migration. I go to Capitol Hill. Jon Beckmann of WCS and I meet regularly with the Upper Green River Cattlemen's Association, which is concerned about the effect of a protected corridor on their cattle.

We illustrate where, how, and when pronghorn migrate. They leave the Tetons in fall and move swiftly, covering about thirty miles a day for three days. In spring, the move from winter ranges northward happens more slowly because animals follow the greening of vegetation, which occurs over the course of a month. Navigating such land means the pronghorn must cross areas owned by private ranchers and managed by different federal jurisdictions—the National Park Service, the U.S. Forest Service, and the U.S. Bureau of Land Management, each with a different public mandate.

The Upper Green River Cattlemen's Association is legitimately concerned that any federal protection might jeopardize their grazing privileges. So we work with them to ensure that hunting, the use of all-terrain vehicles, and cattle drives can continue. Most of these activities occur during the summer anyway—not spring or fall—so to the impact on pronghorn should be small. The ranchers will lose nothing, though we argue that the corridor must have fencing restrictions, with no new structures and no gas fields or leases. Gradually, we build trust with them.

At about this same time, a regional environmental NGO tosses a metaphoric grenade into the discussion, which successfully distracts us from the issue at hand. The NGO, whose mission is to protect grizzly bears and wolves, decides that reduced cattle numbers will lessen conflicts because fewer cattle will be eaten by predators. Therefore, they argue, livestock should be removed. The logic is sound if one is interested exclusively in bears and wolves, but the tactics they use to achieve that goal are far from valid: they opt for a favorite American strategy: litigation.

In addition, their representation of the situation is totally disingenuous. Because they want cattle removed, they use the argument that cattle compete with pronghorn for grazing areas and that the pronghorn would therefore avoid the corridor—a premise not supported by any science.

The threat of a lawsuit is deadly for our conservation efforts as it erodes most local support—never mind that pronghorn are an obvious distraction, with no real merit in the argument for livestock removal. While my friends in the Upper Green River Cattlemen's

Association understand what we have been trying to achieve, many among their more traditional ranching clientele do not. In their eyes, there is no difference between WCS and other NGOs. Even the fumes of Wyoming gas fields permeate once pristine lands, our hopes of establishing a protected corridor vaporize.

We devise a new strategy. Rather than working cooperatively, I'd play contrarian. I ask the cattlemen's association to have us subpoenaed if the case goes to court. The Wildlife Conservation Society would break rank with mainstream NGOs that had been supporting a protected migration corridor. I'd testify on behalf of the cattlemen, noting scientific studies of little dietary overlap between cattle and pronghorn and point to extensive evidence that grass is the mainstay of cattle but represents less than 2 percent of a pronghorn's diet. The case, as it turned out, never reached court, but the threat of it did result in the cattlemen's association gaining an even greater appreciation for the depth of our concerns for pronghorn and for people.

In the end, nearly twenty thousand comments addressed a U.S. government Federal Register notification of a possible policy change in land use, with more than 99 percent favoring corridor protection. In summer 2008, the path was designated the United States' first federally protected corridor. Dirk Kempthorne, secretary of the interior under President George W. Bush, pledged $1 million in support. A strip forty miles long and a mile wide now had security. Still, had it not been for the petroleum industry's immitigable greed for plundering more land, and had it not been for Sublette County's opposition, the zone would have been longer.

Nevertheless, the victory was sweet, and the designation was more than symbolic. A critical path for pronghorn through three bottlenecks was a promise of hope for connecting lands on behalf of all American wildlife. This achievement demonstrated conservation potential when partisan politics disappear. Cowboy boots were not required, but a willingness to listen was, as would be true out in the deserts with saigas and Mongolian herders.

Day 1—Captures, Sharga, Altai Province. I wake early and enjoy coffee in the silent darkness of my tent. The sun rises, spreading its golden mantle across withered grasses. High above, lenticular clouds hug Sutay's dome. Saigas are somewhere on the valley floor.

We've already stretched and anchored the quarter-mile-long net, hiding it in the bosom of a deep swale. The pronghorn mannequins stand sentry beyond the mesh. I instruct the three drivers and their teams: "Take a quick look and see if saigas are within a couple of kilometers of the net. Don't chase, but determine if they are male or female. Remember, we only want females." We're all tense.

Buuvei and Namshir head out in separate vehicles. A motorcycle scout is on his own. I remain at camp a few miles from the net. With Mike the vet, we recheck collars and capture kits. By ten, no one is back. I wonder if the team is finding saigas, whether it's now too warm to chase. An hour later, two vehicles return. They've seen males and females, but the groups are small. The nearest to the net is at least three miles. We'll wait. We play Frisbee and read for the rest of the day.

Day 2—Captures. We head out at first light and drop off assistants to hide at each end of the long net. If an animal becomes entangled, the assistants can be there quickly to extricate them. Everyone in all the vehicles now searches for saigas. I ride with Buuvei. Our Land Cruiser becomes command central. I'll give instructions over two-way radios. The key Mongolian phrases for me are "where," "male or female," "how far from net," "left," "right," and "stop." "Stop," or *zogs* in Mongolian, is the critical word. Each driver must be willing to terminate a chase. Because we'll pursue saigas at high speed, if the chase is prolonged, the animals will surely die. I don't know how long is too long. No one does.

A ranger asks in Mongolian: "What if we kill a saiga?" Another answers: "Bury it, tell no one." I ask Namshir what they've said, and then ask him to translate my words.

"No. If an animal dies, we will tell everyone what happened and why. We cannot hide the facts."

By nine, dust trails the Russian jeep. Ahead of the speeding UAZ are saigas. I make out two animals. They're moving fast. We're alive, alert, focused.

In a swirl, we accelerate and cut decisively. The panicked animals run toward the hidden net, maybe two miles out. We're on two wheels. *Damn.* Back to four. Then, we hit a gully—fully airborne. Back down. *Not good.* My adrenaline pumps.

Our speed increases; sixty kilometers per hour, seventy, eighty-five (about fifty miles per hour). Another bounce. My head smashes against the cab. Buuvei jolts forward. I yell into the radio in Mongolian "males or females?" An answer comes quickly in Mongolian. I understand nothing. Whoops.

I ask Buuvei what was said, but he's now screaming in excitement with one arm taut against the ceiling to avoid more head smashing. We swerve, missing a boulder. The saigas are squeezed between the UAZ and us, about four hundred yards ahead. We angle in. The motorcycle appears to be racing across rocks. Its driver has no radio, and I worry that we'll hit him.

Buuvei is still shouting. I now understand the saigas have horns. Males. *Why the fuck are we chasing males?*

I yell "ZOGS, ZOGS, zogs" into the radio and into the drivers' ears. Finally, we stop, but the UAZ continues. I keep shouting to it. With our vehicle no longer paralleling the fleeing males, they turn abruptly and disappear. The UAZ stops at last. My heart pounds. My head hurts: the ceiling is hard. We all convene at the net.

I ask why they chased males. In the turmoil, people had just forgotten. I remind everyone we want only females. The day is cooler now.

By afternoon, the wind kicks up and clouds arrive. This time, there are six saigas, none of them males, and the net is less than three miles away as the vulture flies. The distance is not good, but if we move animals gently, I'm willing to try. We'll remain a mile distant and move slowly. Everyone agrees.

The animals are unhurried. After about a mile, they are within striking range. We coalesce into a squadron and accelerate to sixty.

With one hand, I hold the door handle; the other presses hard into the ceiling. The saigas cruise at about forty miles per hour. We hit rock, and bounce. Easily outpaced by the saigas on the uneven terrain, we abort.

Just then, the Russian van appears over a ridge, reporting that the saigas are still less than a mile from the net. We dip low between hills and lose the van and saigas. The UAZ does not lose the saigas, though, and is still in the chase. We push hard and spot four saigas. Two veer suddenly away, but the other two see the fake pronghorn and charge toward them.

The net intercepts them: entanglement.

One of our group dashes to the ensnared strugglers, then another of us gets there. One saiga escapes immediately. A moment later, Buuvei is there, then me. Namshir and Mike arrive soon after. We blindfold the frightened female. She's no longer breathing. No pulse. My heart races, *no—really?* *NO. NO.* It can't be.

Then: she exhales. We work quietly to collar, weigh, and release her. The handling time is six minutes. Our adrenaline slows, and we smile. The net is re-erected: we'll try for more.

Day 3—No captures. Saigas are nowhere nearby. *Yesterday's action must have scared away others.* There are a few five or six miles out. The van chases them anyway. I'm annoyed. I thought I had been clear. Later, I ask how long they pursued the animals. Their answers come slowly: "Fifteen minutes, maybe twenty minutes." No one looks at me. *I can't tell fact from fabrication.*

With few prospects for more captures, we'll wait a day. If no saigas appear we'll move to a new site. I remind everyone of our purpose, but perhaps not sternly enough. I repeat again: "Females only."

Day 5—Captures. We isolate females and drive two solitaries at different times into the net, weigh, and release them. All is going well, yet I'm uneasy. Scouts tell Buuvei animals are driven from five miles out, which is just cruel. Thus, credibility remains an issue. I know what I see. I lose confidence in what I cannot.

Day 6, and on—Captures. Two vehicles depart midmorning. I'm the air-traffic controller atop a hill directing field action. The Russian van chases males. Their radio is silent. One pursuit is at least fifteen minutes long and far from the net. This was supposed to be only a scouting mission, so I'm confused, but judge that chase to be at high speed.

Animal harassment is not our goal here. Does it differ from legitimate capture? Am I simply rationalizing my failure to lead?

I'm losing control. Their cowboy antics will kill saigas if they haven't already. Hours pass. I finally make radio contact. We'll all regroup at the net. With measured pauses and deliberation, the translator offers my exacting words.

"I am not happy. For days you ignore my instructions. You chase saigas. You chase males and females for long distances. These are endangered animals. Your government has placed me in charge of captures. Saigas belong to all Mongolians. If you do not want to listen to me, go home. Leave. I don't want you. Find another job."

I walk slowly to the net. I kick the stakes out. In disgust, I throw the net on the ground. Everyone is silent. At camp I eat dinner alone at my tent, not the communal *ger*. Someone says: "Joel is very mad."

Darkness descends. My thoughts grow morbid. Dian Fossey, the gorilla expert, was murdered after years of treating Rwandan locals unpleasantly. The macabre eagle nightmare returns, blood dripping from me. I hope to see the start of a new day; the tent cannot protect me from the sons of Chinggis Khan.

Morning dawns clear. The team apologizes and promises to listen.

Later in our survey, we find a dead saiga. It's fresh, no flies yet. She's four months old. We're fifteen miles from camp. Our on-site necropsy reveals many parasites and prolonged damage to the lungs. The death was not due to us.

Of twenty-two chases, thirteen were unsuccessful, mostly because animals veered from the nets, turned sharply, or outdistanced us on uneven terrain. Of those pursuits that were successful, chase times averaged six minutes, and handling just under seven. Saiga body temperatures increased with chase time. None died within the first two weeks of capture, although a female that had been pursued for eight minutes was dead eighteen days later. She was more than twelve miles out from the net when pursuit began, and while the cause of her death was unknown, several of our other collared animals were chased longer and had higher body temperatures yet survived. Another female lived thirty-five days before being killed by an eagle. The rest lived longer than a year, at which point our collars fell off and we could no longer track the fates of the individual animals.

All eight collared animals moved well beyond the boundaries of Sharga Reserve. One-way movements exceeded forty miles, and those traveling north restricted themselves to a swath less than three miles wide. The Sutay massif limited movements on one side, and the presence of herders and livestock limited it on the other. Each female ranged from twelve to eighteen hundred square miles.

But our work wasn't done. Once the collars were deployed, we arranged to cover the salaries of WWF rangers near the reserve who we trained, along with other locals, to do telemetry—that is, collect data on saiga locations and survival. We purchase a new Chinese motorcycle to facilitate saiga monitoring and, not directly related, raise money from the United States to cover the cost of a surgery in Nepal that is needed by the son of one of our Mongolian rangers. We hope to bolster teamwork and to enhance the local collaborative efforts so important in conservation.

The breadth of saiga movements pales when contrasted to other four-legged Mongolian nomads. The white-tailed gazelles that Kirk Olson studies move over areas ten times as large as that traversed by saigas; khulans roam over an area twice again as large as those traveled by gazelles. Even in such vastness as is found in Mongolia, the accommodation of people and wildlife requires tolerance from both. Given Mongolia's stunning economic transformation from herding to mining, changes to the landscape are rapid. If conservation is going to work and if wild animals are going to persist, the government will need to mitigate its most conspicuous juggernaut—the growth of roads, railways, and poaching.

When I ask my colleagues from the Academy of Sciences about saigas and the likelihood of securing safe passage through the migration corridor we identified, there is silence. Having witnessed the predicament of pronghorn in Yellowstone and having now seen the data on saigas, they understand corridor challenges. They also know the real needs of Mongolian wildlife. Nevertheless, their reaction is unequivocal: "Unless herders want to set aside a specific area, it will never happen. The government will not step in there."

Just the same, we don't quit: over the next three years, Julie Young and I, along with Peter Zahler from WCS and others, instigate a project to look at saiga calf survival on behalf of Mongolia's WCS office. Buuvei will perform the work, which he pursues for his PhD in Massachusetts. More than 120 calves are collared, and collectively we publish a number of journal articles. The peer-reviewed papers help us gain support, but it's mostly the on-the-ground presence that takes precedence at the conservation table and in government circles. This presence ensures that a Mongolian biologist speaks of a Mongolian legacy.

The valuable pashm, or silken underwool, is obtained from these goats,…the animals spending the winter in the highest altitudes, in order that the wool may be as thick and profuse as possible.

CECIL RAWLINS, *The Great Plateau*, 1905

12 Victims of Fashion

It's a cold autumn day as we nudge worn tires over sharpened shale. Soon, a growing coat of desert sand atop the shale protects against puncture. Next come dunes, then mudflats, fissured and cracked. The scant vegetation disappears. Each borehole is dry. The domestic sheep in the back of our Land Cruiser bleats. She smells bad. So do we. She'll become dinner.

Ahead is another well. A camel hair rope and an old bucket hang over it. We crane our necks over the deep, dark shaft and hear silence. I lower the bucket. The splash reveals water that is not illusory. We quench our thirsts and then continue to search for saigas.

Miles farther we stop, scan, and see what we've seen at each stop prior—many goats and sheep. Today we count more than two thousand of these domestic animals. There are also twenty-one camels, more than forty horses, and a whopping three saigas. Snow dusts the high peaks but doesn't reach the valleys. I wonder where the animals find water. Whatever surface water exists must be frozen.

A couple of nomads pace their camels down the mountains; they are the two-humped transporters of their Mongolian existence, a lifeline. Their dogs eye us nervously. We ask how they water their livestock. They must think the question is dumb—let a spigot run to fill a trough. And if ice makes the surface impenetrable, smash it. I understand the tactic. *What about the shy and undoubtedly thirsty saigas?*

Unlike most altitudinal migrants that move down from mountains to avoid deep snow, saigas do the opposite: in winter they go high. There, snow substitutes for water, much as it does for wild yaks. But, here in the Gobi, as elsewhere, humans have become the true arbiters of who survives and how. If people set camp at springs or if livestock concentrate there, wildlife are obliged to go elsewhere. Where?

In these lowland deserts of rocky spires, there are no tall mountains like the Altai in western Mongolia, and there is no reliable catchment for snow. Argalis and ibex are abundant here in the Ikh Nart Reserve, but in years of little precipitation the only reliable water trickles along a couple of denuded streambeds, both of which end after a few hundred yards.

During the winter of 2014–15, each trickle froze. Snow did not fall. Argalis queued above the frozen dribble, and thirty or so waited for days on slopes above the *gers* for a thaw. Regardless of temperatures, these wild sheep need moisture. They approach the ice as if willing the miracle of melt. There was no thaw.

But herders a thousand kilometers east of saiga territory offer optimism for the survival of the argalis. For two days, local people were the doers, breaking the coating of ice with axes and shovels. Their compassion brought water to the argalis. Most animals are not so lucky.

Yet some animals, whether by happenstance or design, have adopted a strategy to help themselves. In this tactic, known as using humans as shields, animals employ people as buffers against nature's threats, essentially making our infrastructures into safeguards. The Scottish-born naturalist John Muir realized this during a 1910 visit to Kenya: "Most of the animals seen today were on the Athi plain, and have learned that the nearer the railroad the safer they are from the attack of either men or lions."

Predation is just one possibility as to why the zebras and giraffes observed by Muir were close to the tracks. It's also possible that the food there was better. Working out the range of possible reasons for some observation and then discerning the facts behind that observation is what science is about. In Ethiopia and Kenya, it didn't take long for researchers to determine that gelada baboons and vervet monkeys spend time near people to avoid the leopards and spotted hyenas that avoid humans.

Examples of human shields being used by animals can also be found in protected areas. For a decade, moose mothers in the Tetons have moved about three hundred feet annually toward paved roads to birth their calves. Given that neither nonpregnant females nor males made such moves, it seems clear that they didn't do this because highways were salted or offered better food. Instead, because bears often avoid roads, the newborns, who can suffer death rates of more than 80 percent as part of the grizzly bears' buffet, gained a reprieve when mothers used highways as a shield.

Five hundred miles to the north of the Grand Tetons is Glacier National Park, where high in the alpine tundra, mountain goats also concentrate near people, though in this case it's not to gain protection from bears. Neither do they do so, as one might suspect, because people feed the goats. Instead, oddly enough, it's because people pee along the trail: because goats are salt deprived, they hone in on the unadulterated urine there. They do so dramatically, leaving their safe haven among cliffs, forsaking the buffer of group living to go find urine, sometimes alone.

If other species make use of human shields, why don't saigas? They did, or at least they tried. In the 1920s, saigas walked into the village of Khovd following a severe *zud*, probably desperate for food.

Their fates are unknown. More recently, near the village of Darvi, saigas grew bolder and approached the village. Like local livestock, they hoped to partake of the food that villagers had put out following a *zud*, but their luck was not good: five were killed by dogs, though others did manage to escape the mongrels. Like most species, saigas were capable of modifying their behavior based on the intensity and variability of their experiences. If they had been tolerated, likely they would have returned and perhaps even grown bold. But a human shield is not something wild saigas benefit from. Having been harassed, why would they return? So instead, they grow shy, not bold, and they do not habituate. Associating with humans became a cause of morbidity for them.

Our counts of wildlife continue. We go deeper into lands baked by sun and scoured by wind. In a sheltered gully there might be wild onion or needle grasses, ephedra or stunted saxaul forests. The places we travel are overgrazed, cold, and barren. Still, we frighten up black-tailed gazelles, red foxes, or a wild cat, here or there. There are hares and hamsters, hedgehogs and jerboas. Yet, the Gobi's most conspicuous denizens are domestic sheep and goats. Our surveys suggest nineteen sheep or goats for every saiga.

Maybe saigas were always sparse. To find out, I examine Russian documents, some as old as fifty years earlier. I talk to Dulamtseren and Amgalan. I ask them about overgrazing or, in other words, competition for food with livestock. Little is known about the ecological dynamics here, despite goats and sheep having had a long history in Mongolia. I read other reports. Nearly everything suggested that saiga ranges had shrunk before their poaching intensified.

Eight centuries ago, in the Chinggis Khan era, there was an extraordinary fifteen-year wet period, known as a pluvial. It had much moisture and little in the way of extreme weather—secrets that were revealed by scientists who measure the growth of tree rings. Some surmise the abundant livestock of that bountiful period led to the initial plundering success of Chinggis Khan and his troops.

Namshir and I decide to find out if the progressive rise of goat and sheep herds accompanied years of abundant moisture. The livestock inspector in Altai Province approves my request to access local weather and animal records. A stenographer soon appears. Her hair is graying, and like contours on a map, her deep facial wrinkles reveal a long desert life. Namshir says something to her, and she goes giddy, her eyes light up, and she steps spryly to another room. She returns with an old dusty journal in one hand. In the other is a remarkable piece of the ancient past, though here, it's simply practical and the only tool available: an abacus, for counting. She also calls out numbers, which Namshir translates to English and I write down. I leave her some extra tögrögs, the local currency, and depart with new records to compare local rainfall with livestock abundance.

Back in UB, I'm not feeling so good. My legs shake and spasm. Maybe it's not even intestinal. Given the scarcity of water for cleaning, I should have been more careful about using antiseptics. There was all that unpasteurized cheese, camel and horse milk, and flies. The handshakes with herders whose hands had been, well, who knows where? I'd been feeling poorly for weeks. The near full bottle of greasy cooking oil dumped in our food every night probably did not help.

At the hotel in UB, I lie writhing on the floor. Near midnight, I decide I need to seek help and go to a clinic. I'm lucky; this week's physician is South African. He spends two hours on me, but can't get a satisfactory diagnosis. So he finds drugs to strip my stomach clean. Three days later, my bowels are normal. I eat again. My legs no longer wobble.

A year passes, and I'm back in Montana. Scientific data stream in from Mongolia and other sites, and not just for saigas but for livestock, khulans, and gazelles from other areas of the Gobi. I contrast figures on the collective biomass of wild ungulates with that for goats, sheep, camels, and horses. Remarkably, even in the protected areas, livestock biomass is nearly twenty times greater than for native ungulates. Irrespective of how I run calculations of precipitation—before, during, or after—goat and sheep numbers keep increasing, even during years of drought. *Zud* years are the only exception: a few million sheep or goats sadly perish during those periods. If herders were more prudent, flock sizes would not be increasing to the point where the land cannot sustain them.

But there is incentive to do otherwise. It's called sustaining a livelihood.

Mongolia and China produce 90 percent of the world's cashmere, a fabric crafted from the fine underhair of goats. About 65 percent of Mongolia's cashmere is sent to Italy, 28 percent to China. The bulk of the remaining textile is marketed in Japan and the United Kingdom. Much of America's comes from Italian imports.

Textiles feature prominently in many cultures. Cashmere is no exception. Its market has been volatile, mixed with the usual ups and downs of supply and demand, coupled with pricing gambits. What was once an extremely popular luxury item—the "pashminas"—and indulged as high-end fashion commanding high prices during the first decade of the twenty-first century, slid toward being considered more commonplace as various entities competed and prices weakened. And that's what I want to know more about: the market's source, its demands, and how it links ecologically to Central Asia.

To learn more, my Mongolian colleague Gana interviews herders on my behalf to discover how profit margins for raw fibers have changed. I examine Mongolia's national consumer price index, which standardizes pricing across time. It's no surprise why goat numbers have increased: money is a great enticement. Herders' profits rose three to five times more rapidly than would be expected as a result of inflation-related growth. Cecil Rawlins already knew this in 1905: "Pashm, though greatly appreciated by the Tibetans for the manufacture of their underclothing, is in such request by the merchants . . . that the demand is always greater than the supply, and the price rises enormously before it reaches its destination." The century-old mantra was still in play—increasing the population of goats increases profits.

I decide to title my talk at 2009 international meetings of the Society of Conservation Biology "Victims of Fashion." Delegates from forty countries will convene in Beijing. The victims referred to in my title are not goats, who are sheered and released. Rather, it's the unintended victims—native ones, like saigas and gazelles and argalis—who suffer when much of the winter fodder disappears into the stomachs of sheep and goats, when domestic herds cluster around available water, or when herding dogs and motorcyclists chase them.

Earlier I had tracked down a series of papers from India's Trans-Himalayan region. The titles told the story: "Perceived Conflicts between Pastoralism and Conservation of the Kiang," "Competition between Domestic Livestock and Wild Bharal," and "Conflicts between Traditional Pastoralism and Conservation of Himalayan Ibex." One scientist consistently on the author list of these publications is Chadrudutt Mishra (Charu), the director of science and conservation for the Snow Leopard Trust. Charu and I agree to have brunch. We had originally met a couple of years earlier in Seattle, but now—after my talk—we discuss the possibility of a joint paper showing how the cashmere trade with the West functions as an incentive for central Asian herders. I ask Buuvei from Mongolia to join our team. Over the next few months, I explore data, including Schaller's, and work with Charu and his team.

With sites spread from Mongolia to the Chang Tang in Tibet, and into India's Trans-Himalayan regions of Ladakh and Spiti, which connect to Tibet, we're ready to unveil our story. Across Mongolia and the highlands of western China and northwestern India, goat numbers have increased for decades to the point where they have come to pose risks to a suite of endangered Asian icons, some of which are well known, while other remain unheralded—underdogs that few people are aware of. The reason for the increase of cloven-footed eaters-of-everything is that herders are rewarded for producing cashmere. The equation is simple: more goats, more profit. It is just as Rawlins said a hundred years ago.

Among the better-known victims of goat-herd expansion are snow leopards and wild yaks, as well as Przewalski's horses and wild Bactrian camels—all of which are threatened or endangered. Lesser known but similarly affected are argalis and ibex, as well as saigas and three species of gazelles. Our study regions reflect common features. Native species are outnumbered by domestics by a magnitude of ten or twenty to one. Asia's protected areas differ in form and sometimes in purpose from those in the West because they also accommodate people. But if reserves are to help sustain native biodiversity, using them as goat factories associated with the garment industry's blindness for profit isn't the best way to go.

In their sponsorship of the cashmere trade, some of the multibillion-dollar conglomerates—though clearly not all—generate effects that reach far beyond domestic animal and wildlife diet overlap. For blue sheep, reproductive rates decline with the flood of goats into their habitats. We didn't know if this was because of reduced food or because they were just forced by fear into areas away from their preferred foraging grounds. We did know that the diets of argalis in Ick Nart overlapped that of domestic goats and sheep by 95 percent; this amounts to a significant problem during winter, when food is limited and domestics are fifteen times more numerous. Also, the dogs of herders kill not just argalis but also chirus and saigas elsewhere, and they drive snow leopards from their kills. With more domestics, there is more predation by snow leopards on livestock and hence more retaliatory killing of the dappled cats. Just the movement of a few temporary *gers* into an area causes gazelles, saigas, and others to abandon areas they had previously inhabited. Disharmony between large native mammals and domestics continues to intensify.

Realizing that Western fashion powers the cashmere trade, Charu, Buuvei, and I write a paper blending economics with ecology to focus on central Asia's wildlife, set within a human milieu. With a dab of optimism, we submit to America's and Britain's finest journals—*Science* and *Nature*. The papers are not even reviewed. We then submit to *Conservation Biology*. Two years pass as three handling editors and eleven reviewers debate the merits and weaknesses of our multiple revisions. It's the way science works, and inevitably it leads to stronger inferences.

Once "Globalization of the Cashmere Market and the Decline of Large Mammals in Central Asia" is published, popular accounts are carried in the *Guardian*, *Asian Scientist*, *Nature in India*, and also in fashion industry logs and blogs in Italy and France. The impacts of fashionable attire on unfashionable, four-legged victims are finally receiving some note. During New York Fashion Week in Manhattan, I place an op-ed in the *New York Daily* titled "The Damage Cashmere Does" and hope to increase interest in this impending conservation crisis; with a different op-ed in *Forbes* I attempt to reach a different sort of readership.

The issue remains thorny, requiring amalgamation of information on economic markets, ecology, and pastoralists, especially those in Central Asia, which harbors much of the world's remaining grasslands. Poaching was once a mediator of changes in wildlife, but our efforts highlight the inadequacy of mere ecological interest to resolve complex human issues. While the growth of traditional pastoralism may also spell the decline of some endemic icons, are these forms of livelihood any less harmonious with nature than our lifestyles in the West? Some argue from cultural perspectives that we should reflexively defend any land-use practice by traditional societies as sustainable, even if evidence is to the contrary.

Biodiversity is on the run. Its conservation can only progress if we identify where the causal problems lie, for without knowledge of the problems, it's impossible to find their solutions. Help from herders, business leaders, and people who understand the garment industry and global trade is critical. Fundamentally, we need to know how to instigate effective societal change while guarding herders from having to pare down what is already a Spartan existence in lands of the high and cold.

For decades, the governments of India, China, and Mongolia have recognized the issues surrounding maintaining healthy rangelands, yet the prevailing views call for blending the lives of animals with those of people. "Blend" and "balance" are the operative words here: though not derived from science, they are steeped in individual or cultural perspectives and often one camp or another claim the moral high ground.

Solutions have been proffered, but these, too, are frail. One possibility is to maintain cashmere as a luxury good, with higher pricing for environmentally friendly sources. Another is to reduce the abundance of goats while increasing the numbers of domestic camels or yaks, which, with their larger mouthparts, would lessen food overlap with smaller native ungulates such as gazelles, saigas, and chirus. Ecologically, therefore, cultivating these big-bodied animals makes sense, and already the fibers of camels and yaks are marketed in Europe and North America. Yet, when Walmart sells cashmere clothing items for twenty bucks a pop, prospects for limiting goat herd sizes appear bleak. Offering some incentives to reduce herd size will be important. Providing access to veterinary services and insurance can also help. There is also growth in business cooperatives—including some like the Patagonia clothing company—that link herders with purchasers that offer compensation based on the quality of the cashmere rather than on weight alone. Herders can therefore sell superior cashmere, which is lighter, rather than increasing herd size to make the same profit.

And, markets change. What is fashionable can rapidly become unfashionable—witness the fall from grace of ostrich feathers in the British Empire a hundred years ago. In the 1990s, information campaigns in India and elsewhere halted the illicit sale of shahtoosh from slaughtered chirus. Green labeling coupled with reduced livestock densities might be a helpful starting point for improved range management, and indeed, pilot projects started by Charu in northwestern India and the WCS in the southern Gobi now explore sustainability issues with herder involvement. These projects, however, like many others, face two serious challenges.

At a broad scale, wearers of cashmere must moderate the desire to be fashionable in favor of sustaining biodiversity. This first challenge will require work with the green side of the garment industry. Coffee pricing and efforts to market the advantages of shade-grown coffee are examples of parallel successes. An example of the second challenge can be found in the driver of an SUV who needs fuel. Does he or she head to the mom-and-pop gas station and absorb a higher per-liter cost in order to acquire a product with a greener environmental impact—or just push the pedal to the next station of a conglomerate to get cheaper pricing? Finally, there are well-intentioned green-labeling efforts that suffer because of unverified claims, compliance, or the requisite enforcement to detect unscrupulous allegations of fair trade or environmental friendliness.

Finding a better path forward will require the fluid integration of planners, seminomads, government officials, and garment industry reps. It will also require anthropologists, economists, and ecologists to share their insights into the problems and their solutions. In the absence of commitment on both global and local scales, the iconic wildlife of the world's highest mountains and greatest steppes will cease to persist. Rather than symbols of success, they will become victims of fashion.

The takin is a strange beast inhabiting a strange country. No animal that I have ever seen is so difficult to describe

HAROLD FRANK WALLACE, *The Big Game of Central and Western China*, 1913

13 In the Valley of Takin

"If you're interested, we really could use guidance with takin," says Tshewang Wangchuk of the Bhutan Foundation. "We don't know much about them, especially their populations. They're scattered about the countryside in deep forests and they migrate high. One of the biggest challenges is that it's a five-day hike over seventeen thousand–foot passes just to get to Tsharijathang."

My eyes grow big. I don't know whether to gulp or smile.

This Himalayan kingdom has tigers and elephants, and a blue ox called gaur. There are eye-popping snow pigeons and sleek clouded leopards, as well as pack-hunting dholes. Leeches are fearless, bears emboldened. There are hallowed black-necked cranes and Darjeeling woodpeckers, four species of hornbill, and even an iridescent pheasant called monal.

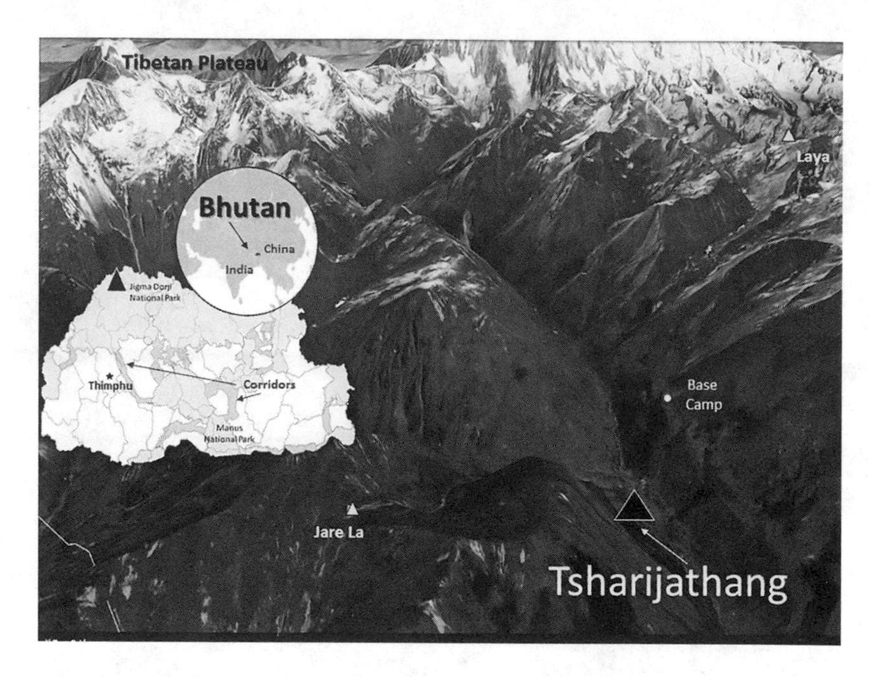

Really: a five-day hike, and snowy passes hidden in clouds. Porters? Yaks? Who has ten days to trek, let alone fly twenty thousand kilometers before starting?

I blurt out three words: "What's a Tsharijathang?"

Tshewang exchanges a quick glance with Tshering Tempa, a PhD student studying tigers. Like two brothers, they know what the other thinks. The room where we are meeting in Montana is silent. I'm not sure of the century; is it the twenty-first or the nineteenth?

My question goes unanswered. There must be a secret here—a *beyul*? A *beyul* is a hidden sanctuary. Tsharijathang must be clothed somewhere deep in the Land of the Thunder Dragon, otherwise known as Bhutan.

Bhutan is as mysterious as it is inaccessible. Limited to high-end tourism, Bhutan does not allow visitors, including biologists, to operate freely. The government has fiercely eschewed the indomitable forces of its two giant neighbors—China and India—and charted its own path.

The land is voiceless, reflecting its violent geological past. Though a mere seventy miles across from south to north, it rises quickly from a few hundred feet along the subtropical Manas River to 23,900 feet at Chomolhari and nearly 25,000 feet at the Gangkhar Puensum—the highest unclimbed mountain in the world—along the Chinese-Tibetan border. Bhutan's other superlatives include being southern Asia's most forested country, with 70 percent tree cover and with 40 percent of its land protected in reserves. Under the strong scientific advocacy of Tshewang, Bhutan has linked protected areas through wide and wild corridors. Whether such actions can protect *Dong Gyem Tsey*—or takin—is uncertain.

This heavily built phantom is the national mammal, though most Bhutanese have never seen takin. According to legend, the spirited Lama Drukpa Kunley—a master of the phallus paintings that still dominate the kingdom's art scene and adorn buildings—deemed the animal a mix of cow and goat. That was five centuries ago.

Isolated in pockets from northern Burma and southwestern China to Assam and western Bhutan, takin have been locally trapped, poached, skinned, and eaten. Their blood is said to warm humans for weeks. Their bones are used in soup, and their hides become decorative rugs. Harold Frank Wallace described their color as "the reincarnation of the Golden Fleece."

The invisibility of the takin is legendary. Veiled by impenetrable woodlands, their habitats overlap with red pandas and macaques. They're obscured by rhododendron and broad-leafed forests of oak and hemlock, hidden by bamboo and banana trees, and vanish behind towering waterfalls on slopes of slime, mud, and rock.

A century ago, Francis Kingdon-Ward reported on one of his journeys:

> Every day the scene grew more savage; the mountains higher and steeper. . . . The great river was plunging down, down,

> boring ever more deeply into the bowel of the earth. Dense jungle surged over the cliffs.... In this paradise we roamed for some time though shivering with cold as the raw wind beat through our garments. Patches of snow still lay melting in the gullies; the mists gathered and dispersed whimsically... Suddenly my attention was diverted by a loud snort, and looking over the ridge I saw on the opposite scree, 300 yards away, a herd of seven takin.

Caught on a tightrope between cats, takin balance precariously: tigers are predators down low and snow leopards play that role up high. Takin escape leech- and insect-riddled lowlands to face domestic yaks and towering summits. And even then, they are not safe from tigers, which here climb higher than thirteen thousand feet. In the Land of Thunder Dragons, the world turns upside down.

Like ghosts dancing among sacred peaks, takin conjure soulful images. In the sheltered hills of Shaanxi, China, they've been called rock goats and, in nearby Gansu, wild oxen. Kingdon-Ward claimed they were "half-goat, half buffalo," and Wallace called the takin the "Rocky Mountain goat." Schaller proffered this description: "Assemble the bulky humped body of a brown bear, the legs of a cow, the broad flat tail of a goat, the knobby horns of a wildebeest, and the black bulging face of a moose with mumps and you have a takin." Others say it's part tapir, part pig. We know takin hindquarters are shorter than the front. We know they live in habitats that are cold, wet, and wild. We don't know all their secrets, though. Counting them can't be easy.

It's a cool February morning along the Mo Chhu in the Himalayan foothills. A cow jingles up a muddied trail, flitting in and out of heavy mist. A rooster crows. I warm alongside a fire of juniper incense. Sipping coffee with me is Tshewang. It's been three years since our planning took hold back in Montana in 2008. A flock of black-chinned yuhinas shoots past; a few moments later a bird with a buffy orange breast and a bluish head appears. Sangay, a government biologist who aspires for a PhD, tells me it's a blue-fronted redstart.

I'm outside Jigme Dorji National Park in Bhutan to help the nascent takin project. The understory is thick with shrubs and bamboo. The canyon walls hide the rugged mountains looming above. Yesterday, rhesus monkeys feasted in a small patch of wild banana. Here, takin cover more ground than monkeys, perhaps fifteen miles between the Gasa Dzong, a tiny communal village, and our field camp atop an open terrace. Across the Mo Chhu, takin demonstrate climbing prowess on vertical terrain. Canopy openings are few and counts impossible. The river is fast and we yell to be heard above its deafening roar.

Our bushwhacking leads to the tracks of wild pigs and deer. At lunch, I wander to the river. Strewn across a boulder are entrails and reddish fur. This macaque was unlucky. A leopard track is nearby.

Sangay, machete in hand, leads the way. Having once studied conflict between tigers and humans, he's keen to locate a suitable graduate program from which to obtain his doctorate and is considering the University of Montana.

If this pilot trip goes well, I'll return during the summer monsoons with an aim to facilitate a study with Sangay. No one knows the status of takin—if populations are stable, growing, or dangerously low. A few days earlier we crossed Dochula Pass, a ten thousand–foot cut stitching its way through the low Himalayas along the forested road to the city of Punakha. It's an area that holds special interest for me, being the location where Schaller had seen takin a decade before.

Estimating populations is always tricky, but the difficulties are magnified here. For one thing, dense cover precludes visibility. And, unlike Mongolia where I could design and lead a project, the Bhutanese are understandably wary of exploitation. Too many times foreigners have come, beat their chests, and disingenuously claimed discovery, including "discovery" of tigers at high elevations. Scientific imperialism has roots everywhere.

I hope to assist Tshewang and Sangay in meeting their conservation and research goals, one of which is to retain the national mammal's best interest at heart in the context of the project. The research must also balance wildlife and humans and not place one over the other and must seek to understand population trends. In addition, the project must provide benefit to the Bhutanese and should neither be simply about testing some arcane hypothesis nor have as its objective the publication of a scientific paper. Threats, including climate modification, must be addressed. The project should have a training component, perhaps even international exchanges with students, but with the understanding that the emphasis must be on Bhutanese students.

Over the next few months, Sangay and I will work on devising a doable plan. Our first glimpses of the needs and reality necessitated by this project will come through the eyes of Tshering Tempa, the biologist who received those brotherly glances from Tshewang in the meeting back in Missoula.

One December, Tempa and I drive to Los Angeles via Death Valley. We camp, laugh, look for wildlife. We discuss research, both takin and tigers. Tempa describes his boyhood, growing up without books and tending animals and crops. Information was handed down through generations by conversation, not reading. Like others in his village of Trashigathang (some 350 miles from Tsharijathang), he had no shoes and walked about in nature barefoot. Science was not a formal topic, and few there had the desire to name many species. Hypotheses were thus unnatural constructs: animals just do what they do.

Similarly, Sangay was not interested in obscure topics—he merely wanted to know where takin go. Their movements were of interest but not the theoretical bases of tortuosity, sinuosity, or Brownian motion models. I reinforced his interest in migration as a dissertation topic but reminded him it would not lead to an understanding of takin population trends.

At my invitation, Sangay visits Montana to meet faculty and other Bhutanese graduate students already there and to explore further the possibility of working on a PhD there. We discuss demography and hike in the Rocky Mountains. We muse about Buddhist approaches to science and try to script his immediate future. I explain why it's critical to know about adult takin survival and juvenile recruitment and that understanding how such vital demographic markers help determine whether a population's trajectory is changing.

If the goal is to recognize whether takin are in danger, we must first know what their current status is. That means applying approaches adopted for species of deep forests to estimate offspring production, as well as occurrences of pregnancy and stress. I ask Sangay to consider capitalizing on the captive takin just outside the capital, Thimphu, and then develop assays to apply to wild takin. We could get a first approximation, at least in Jigme Dorji National Park, if we tag, say, twenty females and use them to estimate vital rates.

Both he and Tshewang had explained earlier that we might be able to count animals and calves on their summer range. The site is a big meadow in a mineral-laden mudflat at thirteen thousand feet, where takin concentrate by the hundreds. It's called Tsharijathang, the magical place accessible only by the five-day hike.

Sangay decides to study migration itself, not population parameters. He'll pursue his doctorate at an Australian university that would require minimal coursework. He'll need to radio-collar animals for his study. These are prospects that enthuse him greatly. He asks that I remain an adviser and participate in the project.

By June, moist warm air will sweep from the Bay of Bengal to the Roof of the World. The monsoons will commence. Along its eastern edge, Bhutan receives 160 to 200 inches of rain each year. That's not my first concern, and neither are the incoming monsoons.

I'll be the oldest person going on the expedition by at least twenty years. To prep for the seventeen thousand–foot passes, I spend the next eight weeks in heavy boots wearing a small pack and running the hills outside Missoula.

Nancy Boedeker, a veterinarian at the Smithsonian who has darted captive takin, will join us in Bhutan. So will Kuenzang Gyelshen, a savvy local animal specialist who has also darted wild takin. Along with Sangay, Tshewang will round out the core team. Nancy and Kuenzang will meet us a week or so later in the field.

By 4:45 in the morning it's light. Unlike the dryness of February, the monsoons pound metal roofs. It's a cool 53°F at Gasa Dzong, eighty-five hundred feet up in forested mountains. A few dogs prowl for scraps, but nothing like the packs of street cleaners back in Thimphu. We mount gear on sixteen wet horses. This includes pots and pans, reed baskets, blue plastic tarps, and even solar panels and heavy twelve-volt batteries. The horses also carry 145 pounds of rice, fermented cheese, and chilies, along with a surfeit of tea and sugar. Tents and sleeping bags are packed separately for transport. There's even a chair for the boss, Sangay.

We depart midday, the air heavy with Bengali moisture. We hike through mud and cross slopes washed clean of trees. We pause in rhododendron stands awash in white flowers. When the clouds lift, the valleys open to our view. Waterfalls plummet from cliffs as if a spigot opens from the heavens. The mud is rich. Gnats attack with the ferocity of Alaskan mosquitoes. We wade through sludge, mostly a combination of pee and horse shit. Sangay wonders when Nancy will arrive. Our team is strung together across narrow mountain trails. By 7:30 at night, it's dark. More mud. We feed the horses. The river thunders below. Logs stack up and break loose to rush dangerously through the current. The rain stops.

My hamstrings are sore and my ankles itch; the latter is the handiwork of leeches. For a brief moment, clouds part. New snow tops an escarpment three thousand feet above. By midmorning the next day, it's raining.

Hours later, drenched and still trekking to Tsharijathang, we reach a military outpost, the last before the Tibetan border. The sun breaks through clouds but soon vanishes behind the gray mist. A friendly dog follows me as I lead the group. It's about an hour up a steep valley to reach Laya, a bucolic village at 11,700 feet in a setting evocative of the Alps. Layap women dress traditionally in bamboo cone hats and thick wool. Prayer flags and yaks abound. The dzong, or fort-monastery, is celebrating a summer takin festival.

The rain has rebooted, and the team converges in a hut to warm. We shed rain-saturated attire near a fire, and I set my boots to dry. Suddenly, conversation halts, and all eyes focus on the foreigner: me. Boots must dry outside. Oops.

The next morning I'm up early. So is our cook, Sonam, who has traipsed along with us here, deep in the forest. He strips wood with his machete and stokes a smoky fire. I offer him coffee. He'll take tea. Today will be easier, with a gentler three thousand–foot gain in elevation. The dog still follows me. It's well taught and knows the difference between Bhutanese rock throwers and me. The temperature drops into the forties; the heavenly spigot still drips.

I ascend a hill and enter a meadow. The dog chases a few marmots. Bad dog, really bad. A massive rock wall waits, and a series of switchbacks cross an earthen dam. Behind is a foreboding glacial-melt lake that grows even larger as I manage a better view. I walk fast, the loose rock and sediment under foot not inspiring confidence. While the chance of this surface rupturing is low, it's not zero.

Bhutan has something like six hundred glaciers and twenty-seven hundred glacial lakes. With continued melting, the frequency of outburst floods from lakes increases. In 1994, the Lugge Tso glacial lake burst. A fusillade of water moved downstream fifty miles, covered fields and farms, and damaged the dzong at Punakha. More than twenty people died. To my west in Uttarakhand, in India, monsoons killed more than five thousand people in 2012 and 2013, as flood waters, mud, and a glacial lake outburst converged.

Gasping, I rise above fifteen thousand feet. There are blue sheep. The dog sees them at the same time, and all them—dogs and sheep—

disappear in a race across scree. I now understand why the Bhutanese throw rocks at dogs. When our group, which was strewn out across a mile or so, reassembles, I express my disgust about the dog chasing the sheep. We speculate that dogs entered the backcountry as a consequence of well-intentioned trekkers who fed the mongrels, who in turn followed them deeper into the remote areas. A similar problem occurs in the Alps, where ravens and crows track hikers for their delectable trash. We all agree to ask the next yak herder en route back to Laya to take the damn dog.

After another high pass, we reach Tsharijathang, the valley of takin. Two rivers and two glacially carved gorges converge. We encounter herders' old yak corrals of rock, collapsed huts, and a stupa. Wild berries abound. Forests of juniper and firs line streams. House-sized boulders polished by past glaciers sit naked in open meadows. Peaks soar to twenty-three thousand feet. Base camp is on an esplanade ten thousand feet below the summits. Sheer walls drop two thousand feet, while glaciers glisten to our north. Tropical air coalesces in the geological drama as monsoonal moisture claws its way up ridgelines. Prayer flags dangle from a lone juniper.

Base camp is soggy. A blue tarp covers our mess hall and shudders in the wind, yet it nevertheless intercepts some rain. I dig moats around my tent. Yak herders come by every few days. One tells us that a two-year-old yak was killed by a snow leopard only forty-eight hours earlier. He preserves some of its meat in a rushing stream and will bring us a leg. Unattended yaks and horses wander about, the lack of tending being the common practice here in Bhutan as is true elsewhere in the Himalaya.

I had thought Tsharijathang was set aside for takin—especially on the rich, mineral laden mudflats. Sangay Clarifies: "It is."

"There are to be no horses or yaks—by agreement with the herders, right?" I press.

"Right."

"But they are there," I say, pointing out what seems obvious.

He shrugs, and walks off.

Half an hour later he returns, ready for action: four of us descend five hundred feet to the meadows to chase the horses and yaks away. The next day, however, they're back. So is the dog, along with two pals. The pattern repeats.

It'll be another week or so until Nancy Boedeker and the vet team arrives with supplies and darting gear. I use the time to feel the land and its rhythms. I rise early, sip coffee, and head out in the gentle light enveloping the highlands.

Sometimes, I set up a scope and watch takin or blue sheep on ridges reaching sixteen thousand feet. Sheep aggregations reach up to sixty in a group. Lammergeyers and griffons soar above. The rain never seems to stop, and I hike to warm myself. When it does stop, and the sun peeks through, I become part thermophile, and like a lizard, I soak warmth from rocks.

Most days, however, it's cloudy and cold. Temperatures hover between 45° and 55°F unless the sun beats down. Neither my tent nor I escape strafing rain and wind. Along the rocky edge of a gorge, I establish a satellite camp. From there, I can spy on takin along wooded slopes a kilometer or more distant, some climbing a thousand feet higher than the Jare La, a pass located at 15,695 feet. It's only when they move through small clearings that I glimpse them and, when lucky, can manage a count. The clouds usually move in from the gorges below and drift back up the valley. Sometimes fog just coalesces. At such times, with no ability to make observations, I escape to a book, climb into my sleeping bag, and drink tea or doze.

Toward evening, the crew and I convene at base camp for dinner—ema-datshi, the traditional mix of rice with red-hot chilies. If we're lucky, there will be a dhal chaser. Following the meal, everyone tells stories, but since English is not one of Bhutan's twenty languages—Dzongkha being the dominant one—I retire to my cozy tent and read. No one else has books, but Sangay is always curious about them, so I lend him a few. This week's is *How Monkeys See the World*, an exposé on hypothesis testing in behavioral ecology.

A century earlier, Ward described the area thus: "In this paradise we roamed for some time though shivering with cold as the raw wind beat through our drenched garments," whereas in 1913, Harold Frank Wallace offered his own version: "There was nothing to do during the day save dry our clothes, read, and stare blankly into the grey wall of mist, which rose grey and forbidding to the very mouth of the cave. Occasionally it parted, and we could see jagged slopes and granite cliffs, with dense bush, far below."

Little has changed—rain, shiver, and clouds engulf one. Often, takin are in the meadows below us, but by the time the veil of mist lifts again, they are gone. I never see takin at the mudflats when horses are present and only a few times when there are yaks.

Given Bhutan's forty thousand domestic yaks and an uncounted number of horses in the northern mountains, I was perplexed as to why wild yaks and kiangs never grazed here. Maybe they had done so historically, but were shot out? This explanation seems unlikely. Neither contemporary nor archaeological signs of their presence have been found. Food is obviously sufficient—otherwise the domestic beasts would not flourish. The nearest wild yaks and kiangs were, until recently, over the Chinese border on the Tibetan Plateau. Recent work in glaciology suggests the most likely answer to their past absence.

Evidence from Holocene glacier fluctuations as well as evidence of Neolithic yak pastoralists both point to the presence of glacial blockages until a few thousand years ago. The situation is similar to the massive ice swells in Greenland that likely obstructed the dispersal of wolves south along the massive Melville Ice Bulge. Here in Bhutan, the entry of domestic yak herders must have been from the north. So, perhaps, just as in understanding why another cold-adapted specialist, the muskox, had not colonized what are now suitable Arctic habitats of northern Quebec or parts of southern Greenland to the south of immense ice bulges, it is geological history that will help unravel ecological mystery. While environments may certainly look similar and indeed may be currently suitable, they have different pasts. In all of the cases just mentioned, there were glacial impediments.

One morning I decide to see if I can watch takin more closely in the forest. My trek drops me down into the mudflats and across a rushing creek where I'm not carried too dangerously far downstream. The climb begins, but the slopes are muddy and steep. Instead of flailing through slime, I huff my way up a trail to Jare La. I slant across decomposed scree, slipping only a little. A musk deer startles me and soon flees.

The clouds descend and the sky reopens its faucet. Forests deepen. The sound of rain on canopy grows loud. There's no visibility, and my heart quickens. Now I worry about tigers.

Sangay told me of one taking a takin above twelve thousand feet. To the north of Bumthang, a herder was killed by a tiger at about nine thousand feet. And to my east, a tiger took a yak above thirteen thousand feet. In his 1899 book *Through Unexplored Asia*, William Jameson Reid described being scratched up by one of these large cats. I need to stop thinking.

Several paths converge, and the trail merges into a takin highway. The odors of fresh dung and urine grow pervasive. I bob and weave below wet branches, trip on mossy rocks, and crawl over roots. My eyes dart sideways, down, and then up. After an hour of more slipping and more mud, I reach an open glade. It's the spot from which I saw seventeen takin yesterday. Today, it's cost me only a tortured five-hour hike—Bhutan's large-billed crows could sail here from my camp in just a minute.

I don't move. Takin neither see me nor (apparently) smell me. It's the time of rut. A male sniffs a cow's urine, retracting his bulging nose and opening his mouth to process her pheromones, an action known as flehmen. Another male stands seventy-five yards out. His neck is thick and his chest appears luxuriantly streaked golden with black. He moves in quickly. The two males display laterally to each other. The newcomer retreats. Some females shuffle among bramble and boulder. I see yet another bull, perhaps two hundred yards off. He sees me, and immediately starts my way.

Cool.

But, I've now lost sight of him. "Cool" was a silly thought.

A century earlier, Wallace described a close encounter between a takin and his best Chinese guide: "Yong had been wounded years before by a takin, which the natives consider a very vicious animal.... With a twist of its head [it] ripped his thigh open. The scar was certainly there." A fellow was killed by a takin as recently as 1999 across the border in India. I also recall how a mountain goat severed the femoral artery of an American hiker. The hiker died. Rupricaprid horns have deathly purpose.

Where is that bloody bull?

My heart is in free fall. I'm soaked by rain, yet sweating. My two trekking poles are in my left hand, a light coat in my right, while a camera and binocular dangle from my neck. The world is silent. *Where is the stud?* Suddenly, I'm realizing this is not a good idea. I inch to the precipice and look down. Standing there—right fucking there!—is the bull. He's looking up, less than thirty feet away. Except, suddenly, we are on even ground.

I amplify my size: trekking poles become rapier horns rising from my head. That's another bad idea.

He wheezes. His tongue flicks. His nasals flare. He moves in, now at the edge. Head sideways, then drooped. Uh oh. I fear lunge. I think thrust. *Not good.*

I toss my jacket. Like a giant eagle swiftly swooping in, it bursts onto Mr. Takin's face.

He spins and, Pegasus-like, launches, one foot hitting vertical rock, as his perfectly muscled body springboards down the cliff face. I refocus. He's now fifty yards out, much of that straight down. I hold out my hand, horizontal. It's not yet shaking.

Throughout, the goat-antelope clan varies in form and lifestyle. While their variations in size may be fascinating, our knowledge of snow oxen is far from robust. The two largest species, muskoxen and takin—Zeus-like in their charisma—command tundra and forest. Yet, we know little of their social styles, even less of their ecological relationships; it is still undetermined how muskoxen shape the lives of lesser species on the snow-vacuumed ridges where they winter, how takin affect bird communities, or the fate of seedlings that return to the soil after passing through their guts. While the group defense of these two related giants may be enhanced by more eyes, more noses, and more ears when in muddied forest glades or on windswept tundra, what of their more diminutive ancestors?

Gorals—or *ghural*, as they are known in the local Indian tongue—are the more primitive ancestral stock of muskoxen and takin. Small in size, gorals live in serious thickets. They escape predators by being cryptic and with ungainly acrobatic maneuvers on cliffs and slopes. They do something else. They have acquired the gift of fearlessness in and around Buddhist monasteries. Akin to their distant relative, the mountain goat, the goral adds human shields to its protective arsenal.

An hour north of Thimphu is the Chagri Dorjeden Monastery. It is remote and white-walled and was established in 1620. From a turnout on a lonely dirt spoor, I reach the monastery by crossing a wooden bridge draped in flags of yellow and blue and then traversing a steeply forested slope. Among the tenets of many Buddhists is reverence for all life. In 1906, Graham Sandberg commented that, "in seeking to account for the vast army of wild quadrupeds in Tibet, another plain reason can be reckoned. With the Buddhist population of Tibet there is ever the strongest repugnance to the slaughter of game [and] while near monasteries no wild creatures whatever are permitted to be molested." A survey of herders up in Qinghai on the Tibetan Plateau more than a century later revealed that 40 percent of them refrained from killing wildlife because it was considered sinful. I'm anxious to see if gorals might be less afraid at Chagri Dorjeden than elsewhere.

My predilection to discover if this is true is flawed unless I'm prepared to compare gorals near and away from monasteries. I'm not.

Yet, I want to watch this lithe progenitor of its more powerful brethren in the snow oxen lineage.

I pass a bleached chorten (Buddhist shrine) and move higher. When I cross the river, four dogs greet me. No owner. They seem innocent enough, I surmise. Shortly thereafter, they disappear. I think not of dogs now, but cats.

In just one small area along the Manas River, only five by five miles, Tempa had camera trapped six species of wild felids. Not only did he find tigers and both marbled and golden cats but leopards as well—clouded and common leopards, and one that carries the formal name of leopard cat. It's a stunning array: 16 percent of the world's felids in an area just the size of Manhattan. The prey in Tempa's photos included gaur and buffalo, four species of monkeys, dholes, and civets. Bhutan's rich biodiversity is bolstered by attitudes of tolerance. I wonder about dogs and human shields, and how species like gorals minimize threats and harassment.

Upon reaching the Chagri Dorjeden, I encounter no monks, only silence. I climb some stairs and circle back to the sounds of chanting. The monks do not understand my quest for gorals. I show them photos. They point to slopes below the monastery walls. Among magnolia and large ferns, I find two gorals feeding. I kneel, but they have no apparent worries—at least not about me. One folds it legs under its body and reclines on a ledge. The other feeds, its ears whirling. A swarm of midges attack, relentlessly assaulting his open nares, ears, and eyes. Both gorals scratch, move, head flick, twist, and turn. I enjoy hours of observation. They are unafraid. I depart not knowing if they are safer because the monastery offers a human shield. We also know little about the responses of gorals to parasites or about how changing weather may affect either the goral or their parasites.

I make my way to an overhang and peer down onto the mudflats of Tsharijathang. Sangay joins me. We are searching for takin. The veterinary crew is a day or so away. Earlier, Nancy had somersaulted down a treacherous section of trail, I find out later, funneling her way to an abrupt ending above the engorged river. Though bruised, she continued. As for Sangay, his study of takin movements will soon begin.

For now, I'm still confused about why the takin are so nervous. A few days earlier, I had encountered dholes in an upper meadow. The mist was thick, and they ran at me, perhaps not knowing what I was. At about fifty yards they swung wide. A herder had once witnessed four or five dholes surround and attack takin; but the dholes charging me all disappeared into the thick forest, and their fates are unknown.

Sangay has not seen interactions between predators and takin but a week earlier had found bear tracks. Most disturbances, at least the ones I was documenting, arose indirectly from the proximity of people—such as the consistent presence of yaks and horses, species that by common agreement were not supposed to be in the mudflats.

From my journal notes:

> The low is 47°F. Fog and rain early. N=29 takin, all in the river basin below base camp; 2 calves, 2 yearlings, 4 adult males. The rest, no clue. 5:20 AM, all run, crossing the river in speedy flight. Calves hurl themselves into the rapids; adults power across. Hmm; why?
>
> 21 blue sheep above 16,000′ A different group of 18 males is about 2 kilometers away. Tashe and Sonam come to my tent and borrow the spotting scope for the day. Dorje borrows my binoculars but promises to return them soon. The old herder returns with the yak leg kept cool in the river. [It had been killed by a snow leopard.] We'll have it for dinner. Kind of cool really. Later, a different herder passing by reports his favorite mule was killed last month. He wasn't sure if it was a common leopard or a snow leopard. His hand is swollen and infected. I provide some medical support.
>
> Three nights ago, dogs were barking, and some food left out at camp is now gone. Yesterday, two dogs passed. They were shy,

> but came back in the dark. I'm guessing it's because we've got too much scattered trash. Guessing here? Yah, right.
>
> July 3rd. The low is 48°F, and the fog thick. We can't see across the valley. The glaciers hide until 9:30 when the sun breaks through. A dull green cover grows vibrant. Gray skies turn blue. One of the scouts sees 27 takin in the mudflats. There are 3 calves. They did not stay long, and split right before five horses came in. It rains. Sitting out under the open cave I'm mostly dry. By afternoon, the gray froth is back. Winds are gale force, and rain is sideways. I'm drenched. I want to keep watching the meadow. It does not feel like 55°. How can I be so miserable in fleece, a hat and gloves? My lips must be blue. I'm shivering. Pathetic, inadequate. I think of Alaska fishermen and Layap herders who tough it out. Wimp [= me].

I wonder about the passive displacement of takin and don't understand their extraordinary sensitivity to disturbance. There is no human shield here, only the antithesis.

Of eighteen days at Tsharijathang, takin appeared on twelve; horses appeared the same twelve days, and domestic yaks on ten. I don't know cause and effect. Takin fled three times when horses arrived and were displaced by yaks directly twice; the other times they remained with yaks but never intermingled with the other species. Herders passed through the mudflats on eight days, which resulted in frenzied and fleeing takin. Free-ranging dogs showed up six times. The number of horses was typically six or so, but once they numbered twenty-five. I consider the additional burden to the animals of our imminent collaring operation.

Nancy and Kuenzang (the local animal specialist) finally arrive. We'll prep for a few days and assign individual responsibilities. We must be ready whenever takin come.

To entice them, Sangay salts an area. Kuenzang builds a blind nearby in order to stalk them at close range. Nancy will dart from a different spot. The rest of us hide on slopes, sitting silently, and waiting for hours. Hours turn to days. It's wet, cold, and foggy.

After a few aborted darting attempts, a gorgeously buff male arrives as the sky darkens. Kuenzang fires anyway. I'm up on a hill and can hear only the roar of the river and a lot of frenzied squawking voices on the two-way radios. Sangay relays in English that the dart hit the rump. The bull speeds off, distancing himself from more possible torment, and disappears into darkness.

Chimidorje, one of our helpers, finds him, slumped over on a steep muddy incline. We work the seven hundred–pound bull under headlamps. His fur is rich, indeed fleece-like, a colorful tan and beige lined with black. He's missing teeth. His abdomen carries a fresh abscess, a gore wound from a rutting rival. We secure the collar. Sangay names him Drukpa Kunley in honor of the person and the sixteenth-century legend who was said to have created the takin—part ox, part goat.

A day later, we dart another. This one crosses the white-lathered river, dangerously swollen by monsoons and melting snows. There is no choice but for us to cross as well. We run along the banks looking for logjams. Tshewang comfortably navigates one without being swept under. I'm next, and nervously shuffle across, whitewater foaming against me. Nancy is next, cautious and hesitant. Time is of the essence because we're unsure where the animal has gone and know only that he's crossed. Chimidorje jumps into the rapids and steadies the logs as Tshewang ushers Nancy to safety on the other side of the torrent. Ninety minutes later, two takin are wearing collars.

In the end, four (two adult males and two adult females) carry GPS technology cinched to their necklines. I'm uneasy about the collaring since one is a young male whose neck will expand with age. Inevitably the collar will press tightly, and then what will happen? Nancy shares my apprehension.

As days turn to weeks and to months, data are slow to come. Signal communication from collar to satellite is blocked because of the thick canopy, and Sangay's concerns about how best to save on battery life prompted us to make the collars active only a few hours daily. We never seem near enough to the animals to detect them, and our data on distribution suffer because we cannot relocate the takin despite their radio collars. But with the lessons we learned this first year, our second year of darting is more effective. Five additional takin are attired by the end of the second year, and then another six the next year. Gradually, the sample increases, and our databases grow. One day, we'll know more about the routes takin take.

Plate 1. Kiangs and decaying Bukubada Glacier, Tibetan-Qinghai Plateau in winter. Slopes beyond the glacier rise to above six thousand meters.

Plate 2. Layap woman with raised arm looks to Himalayan highlands in Bhutan.

Plate 3. Saiga under the Sutay dome, Mongolia's most southern glacier. *Inset*: mother and young (image courtesy of R. Reading).

Plate 4. Muskoxen males (*far left and far right*), adult female with white bonnet, and eleven-month-old.

Plate 5. Base camp at Somnitelnaya on Wrangel Island, Chukotka, Russia.

Plate 6. Snow machines and sledge on Wrangel Island in early daylight; ambient temperature at -19 F°.

Plate 7. Bhutanese wildlife. *Clockwise from top left*: Snow leopard (image courtesy of J. Maier, Wildlife Conservation Society); large hooves and dewclaws of immobilized takin in Tsharijathang; curious adult male takin; blue sheep; mother takin and newborn (image courtesy of Red River Zoological Foundation).

Plate 8. Residents of Tibetan Autonomous Zone.
Top left: Family in background prepares yaks to transport their supplies to lower elevation. *Lower left*: Churning yak butter by hand.
Top right and lower right: Herder families.

Figure 1. Muskox group (*center of photo on sharp ridge*) reflecting their goat-antelope ancestry against wintry backdrop in Cape Krusenstern National Monument, Alaska.

Figure 2. Mother muskox with newborn in Igichuk Hills, Alaska.

Figure 3. Biologist approaches a herd of muskoxen at less than fifty yards in western Alaska to gather photo-imaging data.

Figure 4. *Clockwise from top left*: Model grizzly bear in western Alaska; model polar bear on Wrangel Island, Russia; model caribou on Wrangel Island; two muskox bulls charge (Wrangel Island photos courtesy of Sergey Abarok).

Figure 5. *Clockwise from top left*: Darting muskoxen from air (the two fleeing individuals are already separated from their herd (image courtesy of Marci Johnson); model polar bear on Wrangel Island (image courtesy of Sergey Abarok); polar bears chasing muskoxen (image courtesy of Igor Petrovich); muskoxen charging toward Joel Berger and Olga Starova (image courtesy of Lizza Protas).

Figure 6. Aftermath of wolf predation on muskoxen near Red Dog, Alaska. As seen from the helicopter, the herd waits for three days adjacent to two dead members. *Inset*: the same two deceased females; wolf tracks line the scene.

Figure 7. *Clockwise from top left*: Adult female named Missing in snow hole; Berger approaches Missing; Berger looks down from a snow ledge.

Figure 8. Freddy Goodhope Jr.'s cabin at Espenberg (Bering Land Bridge) on a cold April morning. Harpoon, whalebone, and moose antlers are at left. *Inset*: Freddy.

Figure 9. Wild yak female group on ridgeline adjacent to Bukubada Glacier on Tibetan-Qinghai Plateau in winter, western China.

Figure 10. Wild yak bull on alpine steppe at about 15,500 feet in elevation on a -9°F November morning, Tibetan-Qinghai Plateau in western China.

Figure 11. Two herders above sixteen thousand feet in elevation driving domestic yak in Chang Tang (Tibetan Autonomous Region), western China. Another herd of domestic yaks is visible in the distance, at upper right.

Figure 12. Young chirus in capture facility adjacent to Kekexili National Nature Reserve on Tibetan-Qinghai Plateau. *Inset*: Adult male chiru (image courtesy of Tony Liu).

Figure 13. Curious wolf on Tibetan Plateau. *Inset*: Berger lying in wait as wolf looks on (image courtesy Ellen Cheng).

Figure 14. Hauling supplies across a frozen lake to lighten the load with the jeep still ashore. Windblown loess discolors the glare ice. *Inset*: Jeep subsequently crossing in Kekexili.

Figure 15. A deeply incised glacial valley on trek route to Tsharijathang (Bhutan) with its twenty thousand-foot peaks obscured by clouds. *Inset*: male takin.

Figure 16. *Clockwise from top left*: Petroglyph re-creation in Altai (western Mongolia) of *takhi* (wild horse), yak, and Bactrian camels; mountain goats in Glacier National Park, Montana, United States; Tshewang Wangchuk and Nancy Boedeker crossing flooded glacial torrent on logs in Tsharijathang, Bhutan; goral in Himalayas.

Figure 17. *Clockwise from top left*: Layap yak herders, Bhutan (image courtesy of T. Wangchuk); camels sheared of wool in summer (Altai, Mongolia); impoverished Mongolian family whose herd size is one-sixth that of the wealthier family in image to the left; goats and *ger* with solar panel, satellite dish, and motorcycle, southern Gobi Desert (Mongolia).

Figure 18. Mixed group of takin in Tsharijathang. Two males wear radio collars, and five calves reflect variation in size and birth timing.

Figure 19. Gobi Desert captures of saiga. *Clockwise from top left*: Erecting the flimsy net; model pronghorn; adult female with collar; release of an uncollared young by Enke, Namshir, and Buuvei.

Figure 20. Shearing of goat and underwool in southern Mongolia (image courtesy of Wesley Sarmento).

Figure 21. Lishu Li (*center*) looks for ibex in Ich Nart Nature Reserve, Mongolia. Insets: Lishu (*left*) and Ellen Cheng (*right*).

Figure 22. A male bear approaches a dead muskox in valley bottom. Note muskox herd (black dots) on the white hill above (image courtesy of Olga Starova). *Inset*: female polar bear with two cubs on a different muskox that she most likely preyed.

Figure 23. Images from Wrangel Island. *Clockwise from top left*: Snow machines in snowy wind and fumes partially obscuring a Russian helicopter; early morning coffee in cold cabin; Grisha Nikolaevich and Berger examine photos of muskox-polar bear interactions; a cabin on the island.

Figure 24. Five grizzly bears in Delong Mountains, Alaska. Muskoxen with three newborn babies are on adjacent hill in top center of image (but not visible here). *Inset*: Same grizzly bears.

> In the frozen air, the whole mountain is taut; the silence rings. The [blue] sheep's flanks quake, and the wolves are panting; otherwise, all is still, as if the arrangement of pale shapes held the world together. Then I breathe, and the mountain breathes, setting the world into motion again.
>
> PETER MATTHIESSEN, *The Snow Leopard*, 1978

14

Pavilions Where Snow Dragons Hide

In this highland of cloud and mist, I miss those dabs of brightness that appear in other seasons—the arrival of snow pigeons or a yellowed warbler. Once a vibrant green, the meadows are again muted by monsoonal grays.

This morning dawns uncharacteristically clear as a horizontal sun illuminates the numinous peaks and fresh snows of Tsharijathang. The valleys below will sleep in shadow a few hours more. I peer over a ledge. The only sign of life lower down is a big, black, male yak. My nearest companion is a gray-winged blackbird.

More than a kilometer distant are sixteen takin above tree line. I set my spotting scope. Three calves. The rest are mostly adult. A half dozen more are four hundred feet farther away and closer to crags that drop endlessly below. I'll gather data before the moisture ascends and collides with the coolness of the glaciers. The larger group of animals is spread out, with a couple of adults just standing. Most of them feed.

From out of seemingly nowhere, a musk deer charges toward the group, its whitened side reflecting sunlight. *Odd, a musk deer.* The herd senses something amiss. As if in a choreographed ballet, they clump. Calves are central, almost ringed by adults, a scene suggestive of muskoxen defense. The smaller group flees in a disorganized fashion and stands attentively at the cliffs. The largest group remains vigilant. I'm bewildered. The musk deer is gone. Why the frenzied response to a mere solitary—and far smaller—animal?

Mixed feeding associations are not terribly unusual among ungulates. In Africa, one sees zebras with wildebeests, or springbok with oryx. In the Arctic, I've watched muskoxen with caribou and, occasionally with Dall sheep; in the Tetons and Yellowstone, bison and pronghorn can be seen together.

Tshewang calls in on the radio asking if I've watched the interaction. No big deal I think. He mutters something about the explosiveness of a snow leopard. I ask him to repeat what he, too, had witnessed. Having become inured to the steadfast, boring lives of foraging takin, I'd not made sense of it all. Then, bingo: this was no musk deer with a smidgen of white—it was a snow leopard.

What I had witnessed was the dramatic closure of the spotted cat's predatory attempt on takins. A calf—the obvious target—was nearest the cat, and the response of the herd was instinctive, a formidable cluster of horns and big bodies. With the snow leopard between the two herds, the closest takin fled to cliffs. The larger group clustered at maybe two hundred yards distant, where they did not have the luxury of a vertical escape. With artful skill, Bhutan's national mammal revealed a behavioral response sown by generations of natural selection. Whether taking place in thin air or in deep oceans, evolution has shaped survival tactics for predator and prey alike.

The week before the botched snow leopard attack, I witnessed a different type of relationship unfold. Four free-ranging dogs had eyed me nervously as I was trying to find takin while sitting atop a rocky ledge. The mongrels gave me a wide berth, then descended to the mudflats. After a few customary canine scent markings, they lifted their snouts into the air, having apparently detected an odor, and took off in a synchronous lope. With an eerie slyness, they powered across a torrential current and shook themselves dry.

My single thought was of wolves as I flashbacked to the Tetons, where I watched as well-greased lupine machines readied themselves to pursue elk. But these were feral dogs—large, agile, and powerful. One was mastiff-like in size, another smaller and dark, with a rim of white fur around its face. It took the lead. A third was tannish, and the last dun colored. Somehow, they knew how to configure a cooperative hunt.

The takin were two thousand feet above us and spread about in thickets along cliffs. They need not fear feral scroungers. Or do they? While I had observed a takin group in perfect defense during their snow leopard encounter, both—takin and snow leopards—are native species here with an evolutionary history. Might a novel predator bring something new to the table?

At the top of the crowdfunding site for the Himalayan Mutt Project is the moniker "Pozible," signaling the optimism of the project's founder, Debby Ng.

Debby is a millennial-aged photojournalist from Singapore, a trekker of Nepal, and a graduate student in Tasmania. Her aim is to help people and wildlife. She's trying to do this by attacking a clear nemesis—free-ranging dogs. Despite its stunning backdrops and monasteries adorned in prayer flags of blue and gold, not all is well in Nepal's Manang District. Poverty, dogs, and discord are at play.

Khageshwar Bhattarai, who directs the Himalayan Animal Rescue Trust, reports that the dogs "follow tourists around because they give them food and when they can't find food, they attack livestock and sometimes wildlife." Among the unlucky are blue sheep, gorals, red pandas, and musk deer. Beyond Bhutan, the list of animals that fall prey to wild dogs continues its sad growth.

Debby, like others, wants better canid management. This means reducing conflicts with herders and farmers, minimizing disease transmission, and lowering the uncounted toll on wildlife. Outright killing of dogs is not an option due to the Buddhist belief that all life is sacred. Nevertheless, secret poisonings do occur. Unintended victims can sometimes also include scavengers such as jackals and vultures. The bulk of local efforts centers on reducing the reckless growth of dogs. This means neutering, a neophyte program for which is just a beginning in Manang.

Elsewhere in the world, the dog problem in Nepal is generally out of sight, out of mind. At local scales, it's not.

Once the Tsharijathang pack of dogs I'd observed earlier shakes dry from the river crossing, they vanish. The forest is thick. Three hours later, the sound of barking floats across the upper reaches of the river. I use binoculars to scan distant slopes. Chimidorje is on the radio chattering excitedly in Dzongkha. He's waving his arms a half a mile away. Sangay comes on in English. The dogs have isolated a calf—probably four-months-old and a hundred pounds—from a group. There are no herd mates. I shift to my scope for a better look.

One of the dogs launches himself clumsily at the calf and firmly smacks it on the side. I can't tell if the dog's bite was successful. Another tries to charge the calf's rear, but the youngster is up against a tree. If it runs it's surely dead meat. None of us viewing the scene seems to know where the remaining herd mates have gone.

Over the course of the day and the next one, we watch nine more attacks on takin by dogs. Some target adults. In one, an adult charges three dogs who flee twenty yards and then come at the takin again, and again, and again—different spates, same interaction. The back and forth continues for more than two hours. The dogs have flexibility on their side. They move off, recline, regain strength, and then press in again with more harassment, more attacks, more bites. One time, a takin head butts a dog, which spins off.

Only two dogs are in view toward the end of the second day. Fifteen minutes later, yet another dog pursues a solitary takin a mile away. I'm incapacitated by my distance from the event; there is little I can do.

Novel predators of the dog sort operate in a system where the prey truly are defenseless. Takin did not evolve with dogs, whose very existence here is recent. None of us watching the brutal interactions know the outcome of these attacks. We do know that takin are deterred from feeding when dogs are present, do not remain in the areas where they are normally found, and are spending time and energy warding off dogs.

On a different day, a takin calf wanders alone in the valley. It approaches ten horses, none of which are supposed to be there. They're aggressive, and the calf runs off. Later in the day, the same large, black yak I saw earlier is back. The calf approaches him, probably desperate for companionship, or at least to be with anything other than a dog. The yak is not impressed, threatens the poor tyke, and then cools himself in the river. The calf vanishes. I never know its fate.

A few days pass, and the dogs again reinsert themselves into the lives of more takin at about the time we're ready to continue darting. Three dogs tear off after a dozen takin, and both the herd and the pack splash through a powerful current. A calf and mom become separated from the herd. The mother is valiant in her attempts to protect her investment from the brazen dogs. After forty-five seconds, she abandons her calf, which is now isolated in the cold rushing waters but stands her watery ground in what is basically a safe zone. The same behavior is displayed by desperate elk or moose when they cannot outflee wolves. For the calf, forty terrifying minutes in a glacial stream is anything but good.

At last, we chase the dogs away. The calf makes a final push for something dry and alights on a large boulder. Exhausted, she collapses, the water rushing at her side. We carry the limp animal to shore. Her body temperature is at 96°F (normal is about 101.3°F); she's listless and hypothermic. For a couple of hours, we warm her with our jackets. Nancy rehydrates her with intravenous fluids and vitamins. Gradually, her temperature rises, the shock wears off, and her senses return. She struggles to escape.

She's too small to collar, so we tie a plastic ribbon around her neck before release. That way when, or if, she is sighted again, we'll know that she is different from the other calves isolated by dogs. With no signs of gratitude or a sideways glance, she walks slowly away. For hours, she wanders while uttering groans, probably a distress call for her mother. Toward evening, she disappears in woodlands. The next day she's back, a sign that she survived yesterday's trauma and the night despite dogs and other possible peril. For now, she moves with purpose, back and forth to forest boundaries and throughout the broad meadows. Then, she's gone, never to be sighted by us again. Fate unknown.

More uplifting is that we capture three dogs; one by darting, one by net, and one by enticement. We drag them back to base camp and tie them to rocks and a tree. Each is given a few meager scraps until we figure out what our next steps will be. Someone asks what happens to dogs that harass wildlife in the United States.

"We kill them," I say.

"We can't do that, we're Buddhists," says Sangay.

So for two days, the mutts bake when the sun hits them and soak when it rains. I ask about compassion and an offering of water to drink. Sangay says, no: it's a Bhutanese matter. On day three of internment, they still have been given no water.

I do an end run around Sangay, this time asking Kuenzang about humaneness. His reply: "What, no water?" After that, the mongrels are allowed to quench their thirst. Several days later, they'll be shipped back off to Laya—a few days hike away.

In the end, of the nine takin calves counted, we do not know the fate of three. Did dogs really take out one-third of this summer's crop in just three weeks? And this took place here, in Bhutan's pavilion, a place where happiness is the gross national product?

The scale of interactions between dogs and wildlife transcends the local area. From Tibet to Nepal, and Mongolia to Bhutan, dogs chase, displace, maim, and kill. Victims are small and large, fleet and rare—globally endangered saigas, kiangs, chirus, and wild yaks. Dwellers of forest and cliff—the gorals and serows and takin—are hunted by dogs. The wild relatives of domestic sheep and goats—argalis, ibex, and blue sheep—are also prey. The targets of dogs in India are sambars, chital deer, and blackbucks. In Alaska, it's moose, in Wyoming, elk and deer. At the top of the world, Arctic foxes and muskoxen are harassed by dogs; even polar bears and wolverines are not exempt. Nor are smaller species—black-footed ferrets, salt marsh harvest mice, and kit foxes.

Reptiles and birds are not immune to being plundered by dogs. Along the sandy beaches of Odisha that rim the Bay of Bengal, hatchling olive ridley sea turtles are eaten by them. In California's Mojave, it's desert tortoises. In New Zealand, a single mutt slayed at least six hundred kiwis in just six weeks. Dogs also destroy black-necked cranes and great bustards on the Asian subcontinent. American dogs kill birds small and large—Attwater's prairie chickens in Texas; Hawaii's endemic ducks (koloas), moorhens, and stilts; and sandhill cranes in Mississippi.

The global dog population may number 700 million—about one dog for every eleven people on the planet—with something like 300 million of those considered strays. They harbor disease and infection and are a frequent cause of the displacement and death of other animals.

In areas of low human density, conflicts with wildlife will heighten as people settle and expand, partly because we create situations that favor dogs and other commensals. Our food subsidies maintain dogs where otherwise they wouldn't be. So does our infrastructure. Where sanitation is poor in the Arctic, for example, raven and red fox populations have expanded. Elsewhere, utility poles for power lines offer nesting sites and hunting platforms, effective for the likes of ravens, eagles, and ospreys. It's no surprise that the hatchlings of desert tortoises or sage grouse become tasty crumbs for dogs, given how many of them are poorly supervised by their people. And it's no shock that human provisions enable high dog densities.

Is it likely that prey will be capable of thwarting these new assailants? If they are well-armed with protective devices such as quills or are just large, like elephants, the answer will be yes, and there will be less of an evolutionary impetus to do better in those cases. Otherwise, options narrow, which also seems to happen when climate change is the agent of demise, akin to a race in which an old but steady horse is pitted again a young black steed, who inevitably outpaces the old horse and crosses the finish line first.

History can be a useful guide to gauge the responses of native species to novel predators. Some fifty centuries ago, thylacines (a doglike carnivorous marsupial) and Tasmanian devils inhabited the Australian mainland. They eventually died out there, but both survived on the island of Tasmania into the twentieth century. The devils continue to exist there, but by the 1930s, the large, striped, doglike predators—known also as Tasmanian tigers or Tasmanian wolves—were killed off by humans who were intent on protecting their sheep, which they assumed were being preyed on by the thylacines. The point is that dingoes, which are really pack-hunting survivors, have ancestors that were just dogs. Brought to Australia by Southeast Asian seafarers about five thousand years ago, they colonized the continent in a feral state. Both thylacines and Tasmanian devils went extinct in Australia but persisted in Tasmania, where dingoes never occurred.

Now, it's possible that dingoes were more effective feeders than the slower-paced marsupials and simply outcompeted them for food, rather than having killed them outright. Whatever the cause for the demise of thylacines and devils on the mainland, the bottom line is that newcomers won the race for survival.

The broader view is well known. Native species have ineffective defenses against aliens, a robust example being that of indigenous people on several continents who fared poorly when exposed to new pathogens. If the duration of exposure to these new pathogens is relatively short, and natural selection is intense, then survivors—legged, winged, or scaled—that reproduce rapidly will have offspring that might do all right.

A look at how large kangaroos deal with dingo attacks on mainland Australia is instructive. Roos are discerning combatants who usually but not always flee danger. Males and females differ in terms of both behavior and group sizes. Depending on the species of kangaroo and his or her size, responses vary.

Among grey and red kangaroos, gender accounts for substantial differences in mass, with females being about half the size of males. Males will, at times, fight off attackers. At a place called Wallaby Creek, one observer described male defensive actions that included "high-standing, kicking, thumping the tail, and hopping towards and watching the dingo," whereas another male high tailed it to water. The latter behavior is not all that different from what we witnessed takin doing, which was to seek water when attacked by dogs. In other instances, female roos were seen fleeing longer distances and were more vigilant and flighty than males, all of which makes sense, given their greater investment in the survival of joeys.

Because dingoes and foxes, as well as free-ranging dogs, often concentrate on the young, their predatory actions result in more than just tactics of avoidance by their prey. Around a watering hole in eastern Australia, five dingoes killed more than eighty red kangaroos, of which all but three were juveniles. Such events limit population growth and by no means are restricted to dingoes or marsupials. In Israel, free-roaming dogs lowered the ratio of offspring to adult goitered gazelles to the point of halting population growth. Baikal seals, endemic to Lake Baikal in Russia, received canine distemper from dogs. In Chile, the lithe pudu, an agile member of the deer family, avoids feeding in areas where dogs run free. From Zimbabwe to Ethiopia, more than a dozen species have fallen victim to free-ranging dogs.

The problem is one that won't soon to fade. Tibetan mastiffs figure prominently in the ecology of the hallowed plateau. Twenty-five years ago, pheasants and blue sheep were in the front yards of holy places. In some monasteries now, though, such wildlife is gone because monks do little to reign in the dogs. The gorals I watched were the lucky ones.

In Western cultures, dogs may be our best friends—pampered, well-fed, and recipients of veterinary care—but they are not wildlife's best steward when roaming free. Unattended, they do damage to wildlife.

Resolving these issues in such temperate mountains as in Nepal and Bhutan, or even up in the Arctic, will require compassion for the dogs left adrift in local communities. It will also require teaching people about domestic dog health, performing vaccinations, and suppression of canid fecundity. This will need to be done with sensitivity to individual differences among groups of people and cultural mores, and it won't come about unless public awareness is improved and without, perhaps, changes in laws. Progress will occur due to the work of people like Debby Ng and local NGOs, but other approaches can also work.

Susan Kutz, for instance, has moved beyond her initial work of unlocking the inscrutabilities of emerging Arctic pathogens. Each year she brings veterinary students from the University of Calgary to isolated Inuit villages. Having built trust with the village residents for over twenty years, she leads clinics, gives talks, and has realized that making a difference in controlling dogs requires more than excellent veterinary skills. Whether in the Arctic, the mountain kingdoms of the Himalayas, the towns and *gers* on the Asiatic steppes, or the thorn scrub of Zimbabwe, change will only come about if there is a commitment to stopping the subsidization of dogs' diets by, for example, eliminating trash; it takes a human face.

Nighttime droplets hit the tent. They bead, coalesce into rivulets and race down nylon walls. I hope the shallow moats I've built protect my yellow shelter. The rain increases to powerful bursts. If my sleeping bag is to remain dry, more excavation will be needed tonight.

The logistics so far have mostly worked. Solar panels have charged my twelve-volt motorcycle battery, which in turn powers my digital gear. Cameras are not yet in disrepair, and the cuisine is still spicy. My coffee stash will last another week, as will the powdered milk. Despite the unnerving dog situation and stressful chaos of the most recent darting expedition, no one has been seriously injured. Yak herders have shown interest in our observations of takin and the role of their livestock in disturbing takin populations. I even enjoyed yak meat, an unintended consequence of the kill by a snow leopard. I drift into a fine torpor.

Around midnight, I'm awakened by a ruckus and the banging of pans. Probably just dogs scavenging. I slide back into sleep. The next morning, Sonam can't find his favorite utensils. *Odd—did the dogs carry them away?*

Cordyceps fungi are big business. In summer, collectors trek for days, and cross mountain passes and swollen rivers to pursue their fortune along Tibet's southern border. They come from India and Bhutan, Nepal, and China. They come to capture a fungus that parasitizes the larvae of ghost moths.

The prized fungi or "worms," as it is known in Tibet (from one of its Tibetan names, meaning "summer grass, winter worm"), are sought for trade and marketed in crude local stands or fashionable boutiques from Beijing to Tokyo. They're part of a complicated parasitic organism in high demand for its putative aphrodisiac qualities that purportedly cause better erections or for other herbal benefits. The actual product is a delicate but hard-cased fruiting body, dark and elongate, whose growth unfolds from the head of a dead caterpillar; it is about the length of a wooden matchstick, but three times its width. Collectors scour alpine tundra during the short growing season, hunching over, or sometimes lying prostrate, and staring until their pricey find appears a few centimeters from their noses.

Known scientifically by the cumbersome name—like most Latin epithets—of *Ophiocordyceps sinensis* or *Cordyceps sinensis*, it is only one of hundreds of similar fungi dependent on the insects they infest. Here in the highlands, they're known more colloquially to English speakers as the caterpillar fungus, or more casually, as mentioned above, as the winter worm. In Tibetan, it's called either *yartsa gunbu* or *dbyar rtswa dgor*. Throughout the world of collectors and among a growing number of marketers, it's known by a name that's less of a mouthful—just yarchagumba, or yartsa, for short.

Yartsa is a big deal for two reasons: (1) it's worth a lot of money and (2), as a consequence, it has engendered competition and conflict. All of this has bearing on the regional ecology and wildlife and, ultimately, on the lives of people.

First, the economy: in 2004, the going price for low-quality *Cordyceps* was $1,500 a pound; the high-end versions brought about $9,000. Eight years later, prices topped out at $50,000 per pound, which is comparable to the going rate for rhino horn. Given the rewards, not only do collectors compete with one another but entire villages also vie for access to high-elevation grasslands, where hundreds may gather to search for the pricey prize. The stakes are high. In 2009, seven Nepalese from Manang—the same spot where Debby Ng works—were murdered in efforts to control who has the rights to harvest yartsa. Two years later, nineteen nearby farmers were charged for their deaths.

Second, the ecology: with more people flocking to the uplands, there must be effects. No one has yet looked, but the subsistence lifestyle of many of the gatherers has to have local impacts. Perhaps the collectors are akin to immigrant asylum seekers who flood into the United States, crossing deserts in hopes of a better life. But to get where they are going, people affect the land while traveling across it. The key difference here in the Himalayas is that the amassers of

Cordyceps camp out for weeks or more while asylum seekers tend not to linger in those deserts.

Local governments are peppered with questions about the effects of *Cordyceps* collection occurring on the pasturage of yaks and horses. My Bhutanese partners and I, though sympathetic to human livelihoods, also worry about burgeoning numbers of pilgrims to sensitive montane areas, especially the additional disturbance and displacement they bring to wildlife like takin and snow leopards. There are also inevitable challenges that stem from the additional dogs that often accompany them, for attacks on wildlife by dogs are not infrequent. Local herders from Tsharijathang also worry about the illicit traders, the hungry and undocumented Tibetans or Chinese crossing into their areas that are only a two-day hike away.

Last night's riddle of clanking and the missing cooking supplies is resolved: in fresh mud, two sets of human footprints lead away from camp, and they're not ours. The itinerants who so boldly entered our camp must have been as desperate as dogs—starving and willing to risk detection in order to procure a few food scraps. Thinking of the travelers' apparent desperation, I think also of the natural hazards they encounter—the rivers, snowy mountain passes, and crevasses—and wonder about their fate.

A week goes by with no more dogs and no more thieves. A soldier passes en route to the porous international boundary separating Bhutan from Tibet. He reports finding two bodies in the river. They are together, both Chinese. Collectors of *Cordyceps*?

The trek out will be wet. It'll be nineteen miles across Shinje La Pass to Laya with only a few tall passes, then twenty more miles to Gasa Dzong. I've already clocked some 150 miles at high altitude. A final forty shouldn't be too bad.

Once we reach Laya, however, there are no porters or horses. Plan B requires thirty-six hours to arrange a guide and an agreed upon price. I meet with him at seven the next morning, ready to go, but he throws a wobbly, stating that my duffels are too heavy and that he wants more money. I pay the bandit, and we start for Gasa.

Two days later, soaked and sore, we near the dzong. The trail disappears. The slope is raw, swallowed by mud. Its blanket of trees has been swept aside by the monsoons, which converted soils into muddy torrents; the views are simultaneously breathtaking and intimidating.

I slither across the denuded incline. Each step creates massive suction, with the mud up to my knees. And then there are the rocks and roots concealed in the unstable ooze. When, or even how, the horses—now far behind—will arrive is a mystery. Another helper (Dorje) and I ultimately opt for a different route to Gasa. There, we'll need to find a chainsaw, then return here to deal with the trees that have been swept into the path—otherwise the horses will not make it.

We finally reach the remnants of what had been a road, sloughed and dropping steeply into a gorge. Its replacement is a fierce stream with a waterfall plummeting a hundred feet. The year before, I watched two horses get swept away at a similar swollen crossing. Our river is not dormant.

While not quite Niagara, the deep roar and blinding mist are a good reminder of what would be our fate should this crossing fail. I find two boughs, hand one to Dorje, and keep the other. I give him one of my trekking poles as well. The river is loud. I can't think. Dorje enters the water. His first fifty steps go smoothly. Then he loses his balance. The raging current tears at me. I lumber forth, each step against the savage tide. I ignore Dorje's plight, which is an unneeded distraction, given my own difficulties at the moment. Fortunately, he rights himself.

People colonized the Tibetan Plateau twenty thousand or more years ago. Some Himalayan ecosystems like those in Bhutan were populated only recently when massive deglaciation opened areas only a few hundred years earlier. Before humans walked the heights and long before domestic yaks and horses were on the scene, it was the land of blue sheep and snow leopards, home of takin and tigers. That realm was once untouched by humans, but humans migrated up from lowlands and joined year-round residents like pikas and marmots.

Biological communities are dynamic. Incumbent is the incessant struggle within and between species under an umbrella dictated by climate. These interactions take many forms, such as the conversion of biomaterials into energy or like those between fungi and ghost moths, and all have operated for eons, but at different levels of biological organization.

If there was ever an inkling of ecological stability before the advent of humans and domestic yaks, it fractured with their appearance. That unprecedented dynamic is still spinning, and its grasp extends beyond economy and livelihoods, reaching wildlife. And it bears massively on conservation. Once again, dogs play a pivotal role, with ecological complications that go beyond vicious assaults on native species. Rather, these complications involve zoonotic disease and intermediate hosts, ultimately intertwining the lives of yaks, people, and wildlife.

The disease in question is neurological, affecting the brain and spinal cord, and is known scientifically as coenurosis, or more popularly as gid. The disease process begins when the brain and spinal cord are attacked by parasites that are the larvae of tapeworms that had been residing in the intestines of dogs or wild carnivores. Fifty years ago, it was discovered that dogs associated with the protection of yak herds transmitted gid to young yaks. The chain of interactions, described below, is straightforward enough.

When dogs poop, the eggs of the tapeworms are deposited into pastures. As yaks graze the contaminated grasses, they ingest those eggs, the embryos or larvae of which hitch a ride through the bloodstreams of the yaks, ultimately arriving at their brains or spinal columns. Yaks in the age range of one to three years old are most vulnerable, and up to 9 percent of them die. While snow leopards and other carnivores certainly dealt with the vicissitudes of parasites in the past, as well as such fundamental activities as locating food and finding mates, the arrival of dogs and yaks has seemingly complicated all of that. The extent to which gid and dogs affect snow leopards, takin, and other native species, however, remains a mystery, the resolution of which awaits a next generation of untiring sleuths.

Bhutan today remains at a low human density and is still wild, still remote. But even the country that upholds the principle of gross national happiness is not immune to either outside threats or those of its own creation. Climate modification has put dams in peril of collapse and increased the likelihood of glacial outwash lakes flooding. The growth of human populations affects natural resources, but precautionary measures—such as park and reserve management, changes in forestry and fire policies, and the creation of new educational institutes—can help mitigate those affects. Bhutan designated a national mammal in the 1980s, and now Sangay is trying to develop an effective monitoring program for the takin. Conservation programs in Bhutan already target black-necked cranes, ibises, snow leopards, and tigers.

The fight to conserve what there is and what will remain is always at odds. Like any other, the country encompasses a variety of human lifestyles and desires. In the north, while the number of *Cordyceps* collectors grows annually, the supply of *Cordyceps* does not, inevitably fomenting conflict. Another problem area is the dog situation, which is out of control. And in Tsharijathang, yaks and horses are displacing takin, a species about which much remains unknown, but it's a sure bet that when native species do not possess the capacity for rapid adaptive learning or change, they will lose the race.

In the deserts and mountains between the equator and the North Pole, there are places where people still live at low densities and where some species of wildlife remain in the shadows. If we know less than we should about them—their numbers, what threatens their existence, solutions to the problems they face—it's because their habitats are closed, inaccessible, and dangerous. It's easier to count saigas than takin: the former inhabit open, dry steppe and semiarid grasslands, while the latter are found in forested valleys or rocky, grass-covered alpine zones. Some even more furtive species are still not studied, despite repeated calls from local people and governments. Not everything can be studied due to a shortage of money or a paucity of scientists, and a surprising number of things are as yet unexplored.

There are, nevertheless, bright spots. Bhutan has managed to retain its forests and migration corridors and has created broad protected areas. In 2016, Mongolia set aside their newest nature reserve in the Tost Mountains, a site where snow leopards and ibex make their homes. Trade-offs, though, are the reality, so while some areas will be afforded more protection, others will fuel economic engines through mining and road building, sadly creating winners and losers, economically and biologically.

No longer do I work with rhinos or bison, caribou or moose: such highly recognized animals have numerous advocates. My focus now is on the rare and lesser known. I can't help but wonder what goes into the hierarchy of which animals are championed, which overlooked. Why is it, for instance, that the world knows a species like snow leopards but ignores wild yaks? Many unique and beautiful animals are living at the ecological hard edges along the margins of the world, and yet I wonder why the carnivores among them attract more attention and more money than do the herbivores. Why do the great apes and big cats receive more consideration and dollars than their lesser-sized relatives? I can't imagine that it's really because of their ecological function or adorable looks or because they are shrouded in mystery or exist in abject misery. And while it's true that some populations are precariously small in number, certainly there has to be more to it than numbers alone. Does education hold the key to increased awareness? Is the remoteness of an animal's habitat a factor? Is it a lack of effort by conservationists?

I'm thinking that we—*we* meaning NGOs and academics—have failed to evince greater fascination for diversity itself, for all life forms, and have failed to arouse sufficient passion for the sanctity of life. If people empathized with the remarkably difficult life of a woolly bear moth up in Greenland and its six-year struggle to reach puberty, or with the wood frog in Alaska, whose body tissues turn to ice to survive winter, perhaps then we'd all be doing better. We have no trouble relating to new cuddly kittens or a puppy with soft eyes—short faced and cute, wound up and ready to play. These pets have ample numbers of human advocates. We need to do better for the unknown wild animals, whether the snow ox clan or some other.

I'm reminded of a photograph of a young Jigme Khesar Namgyel Wangchuck proudly holding a young goral. There is a look of pride, a gleam in his eyes matched by a reverence for life, human and wild. In Bhutan, this man known not simply as Jigme but also as his majesty, the king. When a country's royalty and officials appreciate wildlife, there is hope. Similarly, a people who demonstrates respect for the untamed creatures of world give reason to be optimistic about our future. Mahatma Gandhi once said: "The greatness of a nation and its moral progress can be judged by the way its animals are treated." People can demand humane treatment of animals. Animals cannot.

PART IV

Adapt, Move, or Die?

Parents and geography affect who you are, and how you live. The environment helps govern how long that life is. Life: a legacy of land, sea, climate, and chance.

The refrain "adapt, move, or die" has strong underpinnings. If species disperse, they increase their chances of survival. If they do not, or if they cannot adapt, they die. Saigas—whose fossilized bones are found in Beringia and in England—dispersed, and they now survive in Mongolia and Kazakhstan. Mammoths also moved about, but they survive only as fossils now. The populations of woolly mammoths that endured longest were those on a nonglaciated piece of northern Asia, where the East Siberian Sea mixes with the Chukchi Sea. When pharaohs were building pyramids, the world's great Pleistocene elephants survived yet nowhere other than on a desert isle called Wrangel.

Is it silly to think that species can adapt to the continuing climate challenge? The issue at stake in chapters 15–17, and touched on in the prologue, concerns propensity for adaptation and time to adapt, which are contextualized by a single word—“immediacy”—and by knowing whether the pace of human population growth and habitat destruction will outstrip our ability to accomplish conservation.

Ultimately, we need to do more than merely ask questions about what we want for the future. We must also be sufficiently bold to achieve that vision, and we cannot do this without knowing who conservation is for. Is it nothing more than something to make us feel better? Or is it something on a larger scale, such as a biologically diverse global community?

> When we reach the Arctic regions, or snow-capped summits, or absolute deserts, the struggle for life is almost exclusively with the elements.... Not until we reach the extreme confines of life, in the Arctic regions or on the borders of an utter desert, will competition cease.
>
> DARWIN, *The Origin of the Species*, 1859

15

The Struggle for Existence

With these words, Charles Darwin encapsulated the essence of a debate that is still extant. To what extent is life on the fringe governed solely by the purest of environmental forces (cold and heat, drought and snow), rather than those of a biological nature? It's in those extreme geographies where the proclivity to live and reproduce meets the physical world head on. Darwin hardly dwelled on how humans affected wildlife directly. But his writing includes detailed discussions of domestication and hybrids and a wee bit about hunting. He also touched on our destructive path, as illustrated by the letter he cosigned in 1874, urging the British parliament to help limit the cutting of forests on the Aldabra Islands in the Indian Ocean. His hope was to save the great tortoises by maintaining their habitat.

Tortoises there are long lived, just like those from the Galápagos. One on the island of Pinta in the Galápagos, known as George, lived more than a century. But he's best known for his lonely death, having the sad fate of being the last survivor of the Pinta lineage. Prior to all the Pinta females having perished, it would have been possible to rescue these unique tortoises of the Galápagos, but in the end, only George remained.

Much like these reptilian giants, bowhead whales are also long lived. A bowhead harpooned off the Alaskan coast by Inupiat hunters in the 1890s survived for more than a century after, until another harpoon took him down and the sharpened head of the earlier harpoon, found in his neck, revealed the whale's age. Long life is also a trait of the Greenland shark. These matriarchs of the vertebrate world live for four hundred years. Some early-born survivors have literally witnessed unimaginable planetary alterations accumulate from the 1600s until now.

Being long-lived and slow to mature is a disastrous recipe for surviving rapid change, however. A predilection to persist is characteristic of species with high fecundity and robust survivorship—neither are qualities possessed by either large land tortoises or bowhead whales. When the internal biology of extreme species with long generation times confronts the modern human milieu, adaptation is not always possible. In fact, it's usually impossible.

In 1964, the Russian-born geneticist Theodosius Dobzhansky wrote: "Nothing in biology makes sense except in the light of evolution." A half century earlier, the Spanish philosopher George Santayana had suggested that "those who cannot remember the past are condemned to repeat it." Both understood the difficulty of being able to grasp current conditions without an appreciation of what came before.

Escaping biological shackles will be more difficult for many nonhuman species than for us: we have big brains and some flexibility. Yet a large brain is no substitute for being wise. Recognition that we share the same ecosystems as others approaches wisdom, as does learning to avoid past mistakes.

Chance and locale make for interesting anomalies. If humans greatly prize some organ or other body part of an animal, it can spell that species' bad luck, especially when dovetailed with an inopportune geographic distribution. Consider today's rhinos in Africa, which has in common with humans places where biological limitation meets conservation challenge. But first, some background.

Rhinos once lived in the icy Asian Arctic and high on the Tibetan Plateau. Even though the portly giants were coated in thick wooly hair, they went extinct. Today, five species of rhino continue their struggle to survive, including one, the black rhino, that inhabits the world's oldest desert, the Namib. As Darwin noted, extreme forces of pure nature, not just of the Arctic variety, mediate survival.

Black rhinos once spread from the arid wilds of Mali and Turkana across Somalia and throughout lightly forested bush to the Cape of Good Hope. Currently, they persist on the fringe of the vast Skeleton Coast in Namibia. There are no fences. It is hot with little water. Sand-dwelling beetles capture moisture from fog on lee-swept dunes, but rhinos do not, and the average rainfall is a paltry two inches annually. Though Darwin's "struggle for existence" is framed as a biological process where individuals compete for mates and access to food, rhinos have, in addition, adapted to these naked desert conditions.

Well, have they? "Adapted" is an odd word here. Rhinos do what they do. How they do this reveals the intersection between biology and local conditions, and—now—what modern humans have done to muck this up.

Throughout Africa, black rhinos have been reduced by more than 95 percent. In 1989 and 1991, two countries undertook a radical conservation leap. They cut the horns from living rhinos in the hope that hornless pachyderms would no longer be targets for poachers seeking the valuable horns. In Zimbabwe, the operation was performed on white rhinos. In Namibia, it was done on the black.

I went to Namibia in 1991 along with my partner, Carol Cunningham. We lived as nomads in a stark desert, deep with furrowed canyons, few people, and nature's wild brilliance. In the back of our Land Rover was netting, which housed cargo—the most valuable of which was Sonja, our nineteen-month-old. The netting doubled as her bed. Carol and I slept on the car's roof under the Southern Cross. We carried fifty-five gallons of fuel, another forty of water, and all our food. We were dirty and hungry. We lived this way for three years. We wanted to know if the dehorning strategy had biological consequences for the rhinos.

To find out, we needed to understand desert rhythms—relationships between life and death, reproduction and survival—for rhinos living at that environment's edge. It also meant understanding how what happens in one place might be extrapolated to other places. Context matters. Knowing how rhinos live in Kenya or South Africa can improve insights about how rhinos navigate beyond those countries. Food, which is regulated by rainfall, is illustrative: its distribution and abundance affects where and how animals live. We wanted to know rhino home-range sizes to learn about their density in a given habitat and maybe even about reproduction and, perhaps, ultimately, something about horns and survival.

In the parched Namib, individual rhinos used broad areas, with their average home ranges being about 250 square miles. In more productive habitats, like the moister lowlands of Hluhluwe-Umfolozi in South Africa or the Serengeti (before the poaching epidemics), the home range sizes were just 3 percent and 7 percent, respectively, of those in the Namib. Such a striking difference in size—up to thirty times—was magnified by the sparse environmental conditions of the Namib. Those who think of an "average" rhino would be mistaken. A rhino is not just a rhino: geographical locale matters in what defines it.

Rhinos did what they needed to and ranged broadly for food. To find them, Carol, Sonja, and I did the same. Along with Damara tracker Archie Gawuseb—ex-poacher and now conservationist—we tracked about a hundred animals, some in the western deserts and others in Etosha. Without a base camp, we became gypsies. But we grew to know rhinos, including their individual personas and something about their horns.

Calves survived well where mothers had normal horns. We also detected a relationship spanning the sub-Sahara that revealed that, where spotted hyenas lived at higher densities—from Kenya's Aberdare Mountains to South Africa—calves were maimed. Their ears were gnawed and tails shortened or were missing altogether. When hyenas were absent, their body parts remained intact. Where mothers' horns had been removed, calves had poorer survival rates and higher rates of maiming. It turns out that shorter-horned adult rhinos were also subordinated in dominance interactions. In Zimbabwe, where Janet Rachlow did her doctoral research, some 90 percent of the dehorned white rhinos were poached. Dehorning was a valiant attempt to do something, even though it proved ineffective. And it just created another challenge for the rhinos.

Did Africa's rhinos adapt, move, or die? Adapt—certainly not. The horns and their bodies, unchanged for a few million years, worked fine until recently. Today, horns are an unlucky curse. Rhino survival is a gift from local guards, other conservationists, and governments.

In the stunningly dry Namib, rhinos persist alongside desert lions and elephants and mountain zebras and giraffes. As for Darwin's perception about extreme conditions "on the borders of an utter desert" where life might be strangled by the elements, water is the elixir of life.

What about polar deserts? Do they constrain populations at the fringe in the same way as hot ones? Twenty-five years ago, Graeme Caughley and Anne Gunn addressed this very question. They sought to learn how different factors affect caribou, and compared them to red kangaroos, a species of hot deserts. The two species are obviously unrelated by history but are paralleled by the harshness of their respective environments, each with a desert's simple structure.

And each species attains a high local abundance. Both are browsers, and both are prey for pack-hunting canids—wolves in the north and dingoes in the south.

Neither, as Darwin would have surmised, is regulated by predators. For both red kangaroos and caribou, food abundance governed by weather established population limits; this amounted to drought in Australia and growing season length in the Arctic. It was the immediacy of weather changes in both places that mattered most.

For the clan of the snow ox, we know much less. Snow sheep persist at the intersection of mountains and the far north on the Russian side of the Pacific, and on the opposite side of the Pacific, their closest kin, Dall sheep, live. At this union of high elevation and the Arctic, snow sheep survive at eastern Asia's most northern latitudes (nearly seventy degrees north) and above the Arctic Circle. It's where craggy highlands brush against the East Siberian Sea, the closest town being the unforgettable Pevek, a coal-burning wonder that once housed imprisoned dissidents—and me. Beyond what a handful of studies on Dall sheep has reported, we know little of the interaction between weather and population processes in these distal realms.

The same is true at more southern extremes, for, say, ibex in Ethiopia's highlands or, for that matter, diverging north again, for ibex in Siberia's cold interior. Ditto for wild yaks and wild camels, and virtually every species at extreme height on the roof of the world, and others far from there: the impact of weather on such populations as these remains a mystery.

In the absence of critical data, geographical distribution and variation reveal much about adaptive responses. Bears do not hibernate in parts of Mexico, while in Alaska they do. Raccoon dogs hibernate in Russia, but not in Japan.

Species with broad geographic ranges can confront radically different conditions from place to place. There are reasons why bison and pronghorn are no longer mentioned in climate adaptation scenarios, or brown bears, or tigers: all span locales from cold to hot, from high to low. While yaks may not deal ably with warmth, snow leopards prey on ibex low, in the heat of a Gobi summer, and at high elevations in winter in the Karakoram Range. Tigers live in the swamps of Bangladesh's humidity-saturated Sundarbans and in the cold, snowy Sikhote-Alin Mountains in Russia's far eastern Primorsky Krai region. Leopards are similarly found in mountains, savannas, and deserts, from Africa to Asia, including the snowy oak and pine forests south of the Amur River Basin along Russia's Pacific Coast. The bottom line is that we know some species tolerate varied environments, with greatly differing conditions.

What of species with limited ranges under less varied conditions, though? What do their adaptations reflect? A quick digression is needed because the word "adaptation" is tricky and can carry dissimilar meanings.

For a disciple of evolution like Ernst Mayr, "adaptation" meant the products of natural selection. These are essentially evolved forms or structures that enable a species or localized group to persist under a set of unique conditions. The ability of Tibetans to maintain high oxygen levels in their hemoglobin or the brilliantly colored tresses of peacocks are examples. For anthropologists, the meaning is rather different—the behavior of people in area *X* or *Y* might vary. For example, group sizes of hunters in the Kalahari or forests of Africa are viewed as adaptations to specific environments. For scientists and planners examining climate change scenarios, the word "adaptation" is used to prepare business, governance, and infrastructure for future conditions. There is no right or wrong use of the word; it just varies based on one's perspective.

But back to species and distributions. Are we to infer that just because muskoxen, moose, wolverines, and lynx live in northern mountains with cold and snow, they are not adapted for life in snowless warm places? It seems reasonable to assume so, as, for example, neither moose nor muskoxen live in Arizona, where they would overheat.

The world has droves of unique beings, among them, the extremophiles, a suite of organisms living under conditions far more extreme than anything I've ever studied. One such group survives exclusively in the sizzling geothermal vents of Yellowstone: heat-loving microscopic bacteria. These microbes live in acidic conditions that match car batteries and tolerate temperatures to 187°F. The most extreme of these bacteria is from the scintillatingly named genus *Aquificales*, a remarkable, little known organism (*I wonder why*) that is pink, yellow, or white, uses sulfur and hydrogen for energy to support growth, and can be found in mats and filaments in stinky, hot pools. Here's the point: some species do great when the environment is constant. Even black rhinos persisted unchanged for a couple of million years under stable conditions. But when change is rapid, they can't make it. Whether bacteria in hot pools in Yellowstone or pupfish in a Death Valley spring, some organisms can't leave or disperse. So, if the changes are too radical, they don't adapt, they can't move, and so they die.

The year is 2006. It's early winter. The sage is dull, and the sky blue. I'm in the Green River Basin—Wyoming's open spaces with big rigs, manly camps, and plenty of large game. It's where I'm looking for radio-collared pronghorn. Also here are white-tailed jackrabbits.

The stunning hares have molted from their summer grays to a pelage of brilliant white. That creates a camouflage issue in that they don't match what was anticipated to be a wintry background as the big snow is late, and what little has fallen has melted. Zuma, my Aussie blue heeler and best friend, notes the incongruity. She takes off after a mismatched white speedster in the sea of gray sage. Easily outpaced, Zuma returns with one of those doggy quizzical expressions—*how'd I do?* Hmm, yet another mongrel chasing wildlife. Bad dog—no, bad owner.

The more salient question is whether this color mismatch is geographically widespread. If so, might the hares possess the propensity to change biologically and enable a longer future? Once prevalent from California to British Columbia, eastward across the great prairies to Saskatchewan, Minnesota, and Illinois and also hailing from Kansas and New Mexico—essentially western North America above

Arizona—white-tailed jackrabbits have disappeared from a lot of areas. The problem, however, isn't camouflage.

They've vanished from four or five states and provinces. Much of their demise was purposeful: we killed them due to alleged competition with livestock. In so doing, we've once again ignored nature's complexity. With fewer jackrabbits, where do coyotes turn for food? To domestic sheep, at least sometimes. With jackrabbits depressed, where did mosquitoes in California's Central Valley turn for their blood feasts? To us?

Other populations of white-tailed jackrabbits were beset with loss because prairies were plowed. No one knows much about these grassland beauties.

Someone who really understands their relative, the snowshoe hare, is Scott Mills, a vibrant professor at the University of Montana. He and his team have been unraveling how climate affects coat-color variation, its timing, and mismatches. One of their areas of study concentrates on the decrease in snow duration—specifically whether hares possess the ability to adapt the timing of their molt as glaciers recede and snow patterns change. If they don't, and predators gain an upper hand, these forest hares will become a climate casualty.

The snowshoes may have two weapons at their disposal, however, each of which necessitates an ability to alter traits associated with their out-of-sync camouflage. The first is genetic and requires sufficient variability so that survivors of predation not only live but also produce more offspring that survive to reproduce as well. This falls precisely within Darwin's canon of natural selection.

The second weapon, behavior, involves existing plasticity whereby individuals increase their immediate chances for survival by doing things differently. Maybe, when color mismatched, they flee earlier or remain longer but with an element of surprise—like explosive flight—in their toolshed. Maybe they can select habitats differently, choosing microhabitats based on the progression of their molt—all white, all brown, or something in-between.

Marketa Zimova, a PhD student, decided to find out which tools they might use for survival. After years of field tromping and analyses, she reported that the degree of molt mismatch from the background had little effect on their hiding and fleeing. Survival was, however, decreased; when leveraged across short winters, this diminishment in their numbers was significant. Plasticity in existing behavior to molt mismatches was insufficient to combat suggested climate scenarios, so the one positive hope to reverse current vulnerability is through natural selection.

A more immediate and more unruly problem confronts a southern relative of the snowshoe hare, the marsh rabbit. In Florida's Everglades National Park, the newest residents are giant boa constrictors that grow to seventeen feet. The introduced Burmese python is governing the fates of marsh rabbits and nearly a dozen additional park mammals. While the abundance of deer, bobcats, raccoons, and others has dropped by about 90 percent, the marsh rabbits have completely disappeared. So while climate is threatening the longer-term persistence of at least one member of the rabbit-hare family group (snowshoe hares), humans—either directly by shooting them or indirectly by mischief, such as introducing boa constrictors into the environment or through habitat conversion—are posing more immediate challenges.

Some positive news emanates from understanding the intricacies of biology itself. Generation time and fecundity are linked to body size. Species like muskoxen and gorillas have fewer babies in a given year or period of years than do kangaroo rats or coyotes, which are, of course, much smaller in size. The potential for genetic change—hence, for adaptation—is therefore greater in smaller and more rapid breeders. Thus, small mammals like snowshoe hares, ground squirrels, and many others have a good shot at withstanding some change.

For larger mammals, the evolutionary options fade due to long generation times. Evolutionary opportunity doesn't disappear, though, because possibilities for some desirable traits can change quickly over time. We might expect transformations in species under intense selection—some fish are an example, and maybe animals providing the trophies that grace the wall of a hunter's gallery. Wild sheep are often sought by big game enthusiasts—specifically, they want the horns of rams. Poachers also prize the exaggerated traits of natural selection, as in the case of the big tusks of elephants, which fetch a good deal of money. It's relevant to consider whether human exploitation has altered the sizes of horns and tusks.

One perspective on this emerges from the Canadian Rockies where it has always been assumed that wild rams with the largest horns sire the most offspring. It took three decades to figure out that the mating drama among bighorns is not so simple. Marco Festa-Bianchet and Jack Hogg, among others, found that horn growth is tied both to nutrition and genes. It is the role of age, however, that enriches the story. Rams with slow-growing horns have been favored because they do not ascend as quickly to trophy status among hunters and, hence, have more years to be fathers before they become the victims of bullets. Thus, over time, horn size has diminished.

Elsewhere in Canada, more evidence of evolutionary change sown by human harvest has been found. Red foxes come in different color morphs. Silver phases are rarer and so were preferred by trappers because they reaped more money. Across the hundred-year period from 1830 to 1930, the frequency of the silver phase declined from about 16 percent to 5 percent. Switching continents, trophy elephants have been decimated, causing poaching to change from primarily males to small-tusked females. In Zambia's South Luangwa National Park, the proportion of tusk-less elephants increased from 10 percent to 38 percent.

Among the clearest demonstrations of evolutionary change instigated by humans comes from a case in Nebraska. To be a cliff swallow means possessing swiftness and precisely honed wings, both of which enable them to capture insects. A location favored by swallows for nesting and hunting is bridges that cross muddy rivers. A downside is that bridges carry moving cars. When vehicles and swallows meet, there is but one winner.

Biologists capitalized on that unfortunate dynamic to contrast the wing sizes of contest losers splattered on tarmac with a set of random survivors. The intriguing story is how wing size has changed across time. Decades ago, primary feathers were longer, and long-feathered birds were favored for their insect-capture prowess. Today, those with long feathers are the ones smacked more frequently by cars, while birds with shorter primaries are more numerous: the mean size of feathers has shortened in just a couple of decades.

While Darwin's "struggle for existence" concept was steeped in evolutionary thought and not human construct, in today's dialect the idea of species adapting, moving, or dying can be both literal and figurative. Conservation interventions that include a stricter management of humans harvesting animals is what scientists call for. Other practices, less literal and noted below, can create opportunities to replace the philosophy of "do nothing" or "await evolutionary change."

We already know that, in the absence of humans, high-latitude, high-altitude, and low-precipitation sites are places where abiotic extremes such as weather can spell death for the animals living there. Where humans are present and the desire to implement positive change exists, some effects are reversible. The breaking of ice by herders enables Gobi argalis to drink water during freeze-up when snows fail as a water source. The provisioning of food—a kind but controversial endeavor—such as when *zuds* hit Mongolian steppes or when snow is excessively deep in the deer and elk country of America, is another example of intercession by people. Human compassion can make a difference in places like these, but it can also play a role through the maintenance of lands without impediments (e.g., roads and fences), so caribou or saigas can sweep widely during their times of need. The statutory restriction of airplane and helicopter flights across the Chukchi coast does much to help belugas and bowheads from aerial disturbance during migration, and walruses when they haul out onto land. For smaller species less in the eye of the public, such as hares or grasshopper mice or insect-eating birds—that is, those with greater fecundity and shorter generation times than large mammals—adaptive changes will come, albeit slowly.

It's another hot, smoggy summer day in Los Angeles in the 1970s. The Dodgers play at home. Instead of attending the game, I'll go to Zuma Beach, down toward the Ventura County Line. After an eight-mile jog around Chatsworth Reservoir—from which my lungs will need to recover from choking down dirty air—my brothers will join me for the drive to the beach, during which we won't see much besides people.

A few years earlier, I had gone fishing with my Russian-born grandfather amid green valleys in upstate New York. Harry—known as Pop—a dentist and part-time scientist in his own right, told me that only trout from a few clean streams were edible. I asked why pollution was allowed to happen. He answered me with a kind shrug and befuddled smile. Pop did not challenge authority: he felt he had no voice. Growing into adulthood has challenges and we're all shaped by our exposures to people and to the environment.

When in I was in college, I ventured from the LA Basin to the nearby deserts and then the Sierras. I was perplexed by the nomenclature of the area, since Antelope Valley had no pronghorn, and Owl Ridge consisted of a cluster of new homes, with probably not many owls. Furthermore, the state's flag featured a grizzly bear, even though the last of them was killed a half century earlier. Places with kit foxes were now advertised as Serengeti-like, a sad commentary on how we've lost the big and the dangerous but still claim a rudimentary connection to what one might view as broad, wild, and spacious. Wildlife of all kinds seemed in jeopardy during my Californian youth: the last of the condors had disappeared from Mount Pinos, and roadrunners were increasingly rare near the orchards where I picked persimmons, cumquats, and avocados alongside Mexican immigrants. Though it was a tough time for nature, it was nevertheless a good time overall—both social unrest and a quiet tolerance pervaded. It was a period when my mom and a counterculture generation protested a senseless war. People were starting to demand protection of air, water, and wildlife. Music set a new standard. Equality and justice were beginning to be more evenly dispensed.

And nature did still have a few cheery spots. Gray whales migrated back and forth from Baja to Beringia. Desert tortoises were common in the Mojave. Ringtail cats prowled the vertical walls of Topanga Canyon. The American West was relatively unclogged by human development. In my twenties, I had many dreams, among them, working with wild animals in places where there was hope for their enhanced survival.

Today, rewilding has occurred in some of the very denatured places I once tromped. A chunk of the Santa Monica Mountains is now protected under national park statutes. Cougars roam the California southland via passageways established thanks to the science of connectivity. Dolphins and porpoises cruise shorelines. Sea otters extend farther south than where they were in the seventies. Nevertheless, while California has strong environmental legislation, the Golden State grows more crowded, with few prospects of dampening its vibrant population and economy.

When I think of the future beyond California I'm reminded of fascinating evolutionary changes taking place over short time periods. Bill sizes in Galápagos finches alter over years of drought, mustard plants high up in the Rocky Mountains now favor white or pink coloration, disease spreads more rapidly, and changes in immunity to traditional drugs are speedier—these are just a few of the evolutionary changes we know about; still others await discovery. E. O. Wilson famously referred to "the little things that run the world," which could be construed as, basically, a plea to not forget the unsexy creatures—those with short generation times—that create and govern earth's biodiversity.

Despite the role of selection in the size of ram horns mentioned earlier, adaptive change is generally not rapid for large species. Species that are able to maintain plasticity because they are free from imminent habitat destruction, have room to roam, and are not under threat from a gun will enjoy a brighter future. That blue sheep and gorals are able to adjust to living in and around monasteries, perhaps because of their protection by Lamaist monks, signals improved chances of survival, at least when feral dogs are controlled. That pregnant moose shift their habitat to the relative protection that nearby roads provide from bears reflects a sagacity once thought to belong only to monkeys—and humans. That pikas use cool lava tubules in the Columbia River Gorge, where temperatures reach 100°F, is an indication that not all of America's pikas will wilt as the climate warms. A species' ability to vary behavior is the key to its success. However, the fact that black-lipped and other pika varieties are still slaughtered by government sanction in Asia is worrisome. There are places where there will be victories and others where there will be defeats, but the inherent variation within species offers residuals of hope.

Among muskoxen, those near Nome rub against wind turbines even when the powerful blades whoop. Wild yaks learn people are not always dangerous and adjust their use of rarified lands. North American bison enjoy fewer options than do yaks because most no longer roam free, and some 95 percent have spatial restrictions due to fencing. Yaks now have increased opportunities to rove as they once did, as do saigas and wild caribou and takin. But some species need not move just to survive, or otherwise die. Grizzly bears can be behind bars for viewing, just as bison or rhinos are mostly fenced. But, if we can adapt our attitudes and accept some of these large species, their prospects to live more fully on unfettered landscapes improve.

The future for these sorts of animals is complicated. Though we may naively believe that all we need to do to safeguard animals is protect the land they inhabit, species survival requirements differ, and governments may not have much flexibility to implement additional protections until more people demand it. The cashmere trade issue, for instance, needs good minds working on it and actionable conservation if a dozen or so of Asia's iconic wildlife species are to have a healthier future. Currently, the issues confronting takin are not about genomes and probably aren't about climate change at this point, either. We still don't even know their population sizes and vulnerabilities. Free-roaming dogs, however, are clearly not helping them or their snow oxen relatives.

If we can't count on evolutionary change to help the long-lived escape their biological history, what can we count on? For them, what does winning the race for survival look like?

> If life within the Arctic Circle were perfect comfort, everybody would be coming here.
>
> GEORGE DE LONG, *The Voyage of the Jeannette* (whose ship was later crushed by the ice), 1879

16 A Postapocalyptic World—Vrangel

Asia's claim within Beringia is a forgotten section of the planet that is nine time zones from Moscow. Its mountains and tundra harbor few people. Stretching from the massive Kolyma River to the Kamchatka Peninsula, and from the Sea of Okhotsk and Pacific Ocean to the Arctic Ocean, Beringia encompasses the Chukchi Peninsula and much of the land to its west; in the northern part of it are a few nonglaciated islands. Russia's easternmost sovereignty ends at a volcanic plug called Big Diomede Island, only twenty-eight miles from the Inupiat hamlet of Wales on American tundra.

A different tract of land, surrounded by a different frozen sea, has become my beacon. Called Vrangel Ostrov in Russian, it's better known to much of the world as Wrangel Island. It's the Arctic's only World Heritage Site, a prestige bestowed on it in 2004 by the United Nations through UNESCO. To reach Asia's Beringia from America is difficult.

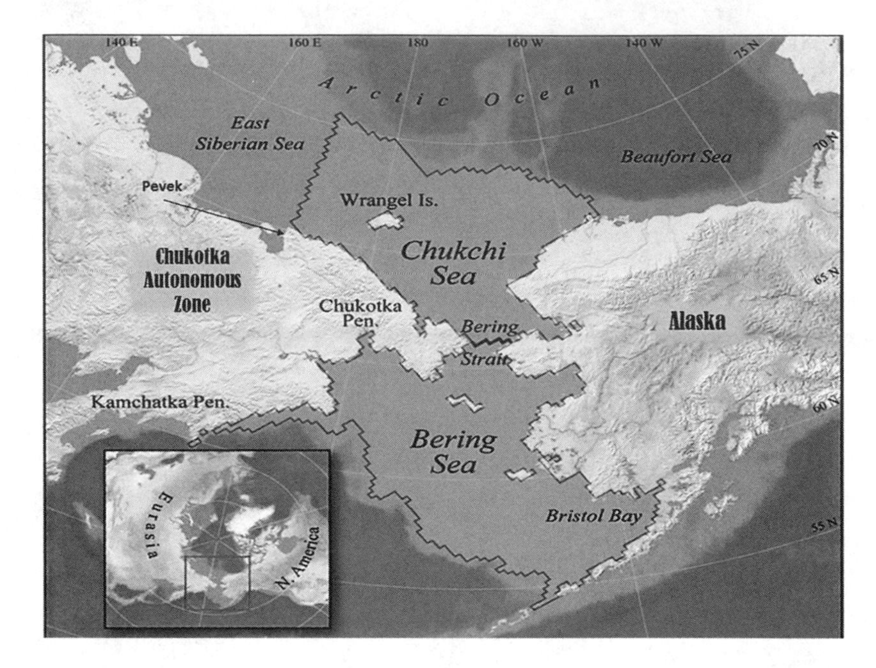

No longer can one just hop across from Alaska. Politics and money dictate that travelers must now circumnavigate the world.

It's a wintry Wednesday when I board a Yakutia Airlines flight in Moscow with two trunks and four extra-large duffels, leaving in darkness and arriving in a foggy daylight soup of icy precipitation. The jet skids to a stop in Pevek, but touching down on solid ground is a welcome event nevertheless, despite an outside temp of −22°F. I grab my pack and get ready to exit.

Russian Security Forces board, looking for me: the Americanski.

"Passport?" they demand.

As soon as I produce it, it's confiscated. I'm to report to the police in two hours, which I do, and I'm officially detained. My court appearance and sentencing are set for Saturday. I'll dress in my finest Carhartts.

The magic day arrives. A picture of Vladimir Putin hangs above. I try to mirror his serious expression: lips puffed, stare stern, eyes forward. The only non-Russian expression I recognize is "CIA."

I'm welcomed to Chukotka's Autonomous Zone.

A single word sums up the Mad Max films: "apocalyptic." Lawlessness and barbarity blanket the world of Mad Max. We're not quite there yet, ourselves, but the plundering humans has done has still been plenty, from countless disasters at sea and on land, from Bhopal to Fukushima, and so far beyond.

Ecologically, this continues. Alien species—Africanized bees, zebra mussels, and kudzu—leave chronic signatures on the new habitats into which they're introduced. Even after nonnative rats and cats have been removed from islands where they were introduced, the rebound is slow. Fecund generalists—cockroaches, coyotes, rats, and raccoons—are thriving in cities. Miami's $24 billion tourist industry has been negatively affected by fears of the Zika virus, while 175 miles of the Yellowstone River were closed to try to reduce the impact of a fish kill caused by an invasive parasite, threatening Montana's $1 billion tourism machine. Most of us may be detached from nature but we are not exempt from the disharmony that has been brought on within it.

Nonetheless, there is hope. People still love nature and animals. Optimism should flourish in the vast realms free of human meddling.

A place of polar perfection must be Wrangel Island: remote, three hundred miles above the Arctic Circle, and isolated from land by eighty miles. Pevek, the nearest village, is 240 miles away. The island's interior plateaus get as warm as 90°F. Virtually free of past glaciation, the island has peaks that jut to three thousand feet. Wrangel has Asia's highest density of reproductive polar bears, two species of lemmings found nowhere else, and the continent's only population of snow geese—which winters in two widely separate locations, the first being temperate Washington State and the other, sunny California. There are polar foxes and snowy owls. More than a hundred migratory birds nest in its wetlands and on its vertical cliffs. Gray whales and belugas feed in its protected waters, along with a hundred thousand walruses. Twenty-three species of plants are endemic.

The roadless treasure is about eighty miles long and half as wide. It'd be easy to get lost here. The earliest ground description was written in 1881 by John Muir—yes, that Muir of Sierra Club fame. That same Muir who explored Alaska's glaciers and later commented on human shields in Kenya in 1910.

Wrangel became a state reserve within Russia's protected area network in 1975. That same year, I started a PhD program at the University of Colorado. My thoughts were not on lonely Arctic lands, a slanting sun, or the sweep of a polar night exploding in florescent green. Neither were they on the spectral glow of a hallowed red sun, pressure ridges of cobalt ice, or misty fjords. Wildlife species accustomed to icy grandeur were nowhere in the high country beyond Boulder.

Three years earlier, when Richard Nixon had made his historic visit to China, polar mammals were not in his thoughts, either. But because international diplomacy can be creatively packaged in the form of symbolic animals, the Chinese gifted Nixon two black-and-white, yet nevertheless colorful, jostling pandas. Whether such kindly tokens of peace benefit the species themselves is arguable, but such acts do attract attention to the species and gobs of money to its conservation. Nixon's quid pro quo was special as well, involving an animal whose babies could easily outrival baby pandas for cuteness: two muskoxen were delivered to the Beijing Zoo.

Cooperation and hope follow unpredictable paths. The year following Nixon's resignation as president, the U.S. government again bestowed an animal favor on a foreign nation. Though the recipient was different, the species was the same, and in 1975, twenty muskoxen from Alaska arrived on Wrangel Island.

Almost forty years later, that population of shaggy-haired beasts would bring me to Russia, despite the fact that, in 2014, Crimea had just detonated, owing to Russia's seizure of it from Ukraine. The tone of the U.S. reaction to the annexation was brusque, conveying a deserved, cold harshness. Meanwhile, in Montana, it's −3°F on a snowy February morning as I board a flight, the first of several that will take me to Moscow, where I'll connect with the Yakutia jet that will carry me farther east.

In the interest of collaboration, my work is sanctioned by both governments. Obtaining the permits for research, for travel into sensitive border zones, for entry into different parks, and even for presentations on science to the public requires patience and more than a year of planning. Russia, for its part, wants to make certain that I won't embarrass them the way Eric Snowden, the American ex-pat living in Russia, had humiliated the United States. My sponsor is Alexander Gruzdev, a big bear of a man with a nice smile, a throaty Russian accent, a PhD in biology, and the ability to knock back a lot of vodka. Alexander is also the director of the Wrangel Island Reserve.

Following interrogation by three KGB-type groups, Alexander offers assurances that I am not political and not an agitator and garners my clearance to help Russian scientists. Together, we plan to develop an ecological baseline from which to assess climate and other impacts to Wrangel Island. The year prior, I had sponsored Alexander's trip to Montana for training in photogrammetry, which would be our chief tool in the study of our target species now on Wrangel: muskoxen.

Wrangel Island carries a certain profoundness, a divine story of a different sort. Life here is postapocalyptic, a kind of rebound. It portends the future. I'll get to see it now.

Alexander, who otherwise goes by Sasha, is not the best communicator. On the occasions when Sasha tries to speak English—which I appreciate—much of his meaning is lost and so I don't know what to expect on Wrangel.

One thing I do know is that, as part of the forgotten Russia east of the Urals, it will be primitive. I'll be the Americanski in a beefy culture where tanks and testosterone rule. It's a place where songs blare "I will drink your blood," and where a large poster of a topless woman is shamelessly displayed on the office wall of Wrangel Reserve's deputy chief. Russians who've never visited the States hold Americans in little regard, assuming that they're unable to hold their liquor, hike up a mountain, or skin out a deer.

Given Sasha's sparse attention to details during the planning stages of my trip here, I pack two warm sleeping bags and three face masks. I shove an assortment of gloves, goggles, a massive parka, and my requisite coffee, along with three pairs of boots, one pair weighing eight pounds by itself, into a duffel. Into another goes muesli, dried currants, power bars, and thirty-five freeze-dried meals. I bring along a twelve-pound Inmarsat system for communications, emergency flares, and two-way radios. I also pack fifteen pounds of batteries. Fares for exceeding baggage weight limits come to forty dollars a pound. I've made the airlines happy.

After my third security check in Pevek, I board an Mi-8 helicopter, which lifts off with a deafening whirl as blue and black exhaust strews across its darkly stained bow. Sitting beside me are four naked bodies, female and male: these reindeer, frozen solid and skinned, along with an equally dead white hare, offer a clue about Sasha's food arrangements.

In the two decades that have passed since my last fieldwork in Russia, time here has stood still. Huge drums of gas are cabled to the inside of the helicopter next to the reindeer and me. Choking fumes move from the outside in. Chains dangle from the hull, a concoction to ensure our rotors aerodynamic efficiency. Seat belts are nonexistent, and there is no sound abatement. There are no rules. It's cold and drafty and more miserable than exciting.

Prior to takeoff, there had been a heated shouting match on the runway, of which I understood nothing. Yelizaveta Protas—who is part Russian and part American, with fiery eyes and a beguiling smile, possesses both language fluency and calm, and will be my interpreter—explains, now that we're finally in the air, that the uproar had to do with the fact that we are permitted slightly more than two metric tons of cargo, and we have three. The pilot had refused to take off, but his inclination to fly grew increasingly more positive with the promise of money discretely finding its way to him. Now, thirty minutes into the midday flight, Yuri and Mikhail open the vodka.

Three hours later we're on the island. Two dilapidated snow machines greet us, their drivers in thick fur, wool, and camo. We haul the animal cadavers and our cargo to a cabin and spend two days making Somnitelnaya functional as our temporary base camp. Snow will be melted or, more traditionally, stored in rusted fifty-five gallon drums for drinking and cleaning. Wooden scraps, the leftovers from a bewildering array of ramshackle shelters, will be used to build a fire and melt it. Scattered among the buildings are rotting pieces of machinery and other Soviet remnants, some from when the United States and Russia were real allies—that is, during World War II. The outdoor commode is filled with eight feet of drifted snow and so, instead of finding shelter in times of biological need, I will meet only wind.

Our cabin is a sturdy old beast. Snow reaches the roof and blocks light from entering windows. Rusted spikes on the windows deter polar bear entry. The Russian flag flaps above. I'm eager to start my investigations, but help the team shovel, saw pieces of ice for water, and haul chunks of wood containing corroded nails to build a fire.

In Russian, muskoxen are called *ovtsebyk*, which literally means sheep and ox. This is not a bad interpretation of their true biology, except that muskoxen have no ox in them. I use the drifted snow to climb to the high point on the cabin's roof. Black dots are in the distance—muskoxen.

Inside, the atmosphere is jovial. A choking haze of nicotine permeates the room, while a corroded stove pumps charcoal plumes sideways. Vodka has been out all day. Dinner arrives, which simply means that tins are opened and their savory wares are speared. More smoking and more vodka. Music from Moscow shrieks. A huge vat of drinking water boils away as darkened shards float from the ceiling into it. *Lead or asbestos?* Perhaps I'm the only one who notices. I retire early.

By four I'm up. Sasha and others have been at it all night. They offer vodka. I head to the outside bathroom in cold and darkness.

Wrangel's weather is frigid, but warmer than Siberia's interior, where it can drop to −80°F. This I know because the island's meteorological station has operated since 1926. Six times daily, workers check wind, temperature, precipitation, and humidity.

Nonnatives knew nothing, not even the location of this conjectural land, midway into the nineteenth century. As early as 1822, Chukchi natives had told Ferdinand Petrovich Vrangel of a wild country far to the north, with tall, snowy mountains. Vrangel failed to reach the island, as did American whaler Thomas Long, who nevertheless named it Wrangel in 1867, the year, as it happens, that Alaska was purchased.

A minority belief then was that the ocean at the top of the world held warm water, fueled in part by the putatively gentle Kuroshio current out of Japan and the Atlantic's Gulf Stream, a fusion sufficient to take the ice out of polar. The speculation was so great that, in the nineteenth century, Captain Benjamin Franklin Homan rhapsodized: "How beautiful and warm and pleasant it will be in that warm sea around the north pole whare thare will be found all sorts of life and sumer fruits."

In 1879, Captain George De Long commanded the U.S. Arctic Expedition to discover the truth. Their vessel, the *Jeannette*, carrying thirty-two sailors, was frozen in pack ice and drifted on currents for twenty months northwest of Wrangel and, in the end, was pulverized. Delong and the other crewmen died, despite an intrepid thousand-mile trek across sea ice to the Siberian mainland.

The first to overwinter on Wrangel didn't fare much better. In 1913, Captain Robert Bartlett led the Canadian Arctic Expedition. Bartlett's ship, the *Karluk*, was also destroyed by ice. The five-person staff rescued what they could from the ship, but their valiant attempt to endure Wrangel's permafrost was not very successful. Only an Inupiat woman from Nome named Ada Blackjack survived. Bartlett had earlier gone for help, crossing pressure ridges to reach the Siberian mainland. Eventually, he caught a ship to Alaska and sent a rescue vehicle for Blackjack.

I'm being jarred about in a wooden box on runners, the Russian version of a sled, pulled along by a primitive snow machine as we scour mountains and river valleys for muskoxen. The sled actually works well, except that I'm reminded that sitting still when the standing temperature is −20°F is a bad idea. The feeling is worse when sitting in wind chill of −45°F. My feet—especially my heels—are quite cold, but the Russians seem never to complain, so I say nothing.

Numbness spreads to my fingers. I take heart that my goggles—only half frozen by inner fog—allow me to see half the landscape. A wolf pack has been traveling past, and I wonder what I've missed in the foggy other half. The wolves obviously crossed a lot of pack ice to get here. *Is their major prey the island's reindeer or muskoxen?*

Before glacial melting led to sea-level rises, Wrangel connected with the mainland as part of the vestigial bridge joining Asia to North America. We bounce over tundra where lichens and grasses pierce a mantle of snow. I think: African plains. Though here there would have been saigas instead of gazelles, bison and not Cape buffalo, wild horses and no zebras. Wooly mammoth, no elephant.

Among our group is Olga Starova, the reserve's twenty-seven-year-old lead scientist. Motivated and fit, she's equally game to walk ten miles or bounce on the back of the carriage that now drags us and Yelizaveta Protas, or Lizza as she is generally known. The only problem is that Olga knew nothing about this project before I arrived. Sasha, the great communicator, had not mentioned it.

She's keen to learn photogrammetry. And our sturdy guide is intent on keeping alive our badly beaten Buran—an early version of a snow machine that is now wheezing. It's the snow equivalent of a

1970s Honda 50 dirt bike with no guts. It breaks often yet assures the luxury of getting one field day out of every four days.

Today we abandon the stalled Buran and summit a thousand-foot high ice dome to scan for muskoxen. We step over wolverine and polar bear tracks. No one knows how many bears are on Wrangel. A week earlier, we saw three mothers, each followed by a pair of cubs. Some bears are more than twenty miles inland. A 2004 survey tallied 261 bears. I counted more than twenty in a single tundra patch clustered above bird colonies in a 2011 photo. That scene was reminiscent of whalers' reports from the 1870s of high bear numbers on Saint Mathew Island, south of the Bering Straits. In 2017, Alexander Gruzdev snapped photos of more than 240 bears on land near a whale carcass that washed ashore.

Bartlett expedition members on Wrangel were understandably troubled by the white carnivores. Ada Blackjack wrote: "Polar bear and one Cub very close to the Camp and I didn't take any chances." The Russian Pomors, mentioned in chapter 3, did take chances: stranded for four years on Svalbard in the 1600s, they speared the white bears with nails mounted to wood. Why would we ever worry about ice bears now? We have a single modern weapon, a torch that shoots flames fifteen long inches.

One early morning, I enjoy some calm before our cabin's crowd bustles. I depart for bathroom duty. The wind is blowing, and it's about zero degrees out. My exposed butt grows numb, but finally I cinch up my pants and head back.

Two sets of bear tracks cross those I made en route here from the cabin. *Damn, that was just moments ago.* In a muted panic, I survey the area, but the only thing I see moving is a waft of snow. *Two bears, where?* I quickly scan nearby debris for something to climb—heaps of wood, a tower, an old tank, shipping crates, a rusted container, anything.

Nothing.

It's 150 yards across no-man's-land back to the cabin. My heart races in high gear. I reach the cabin. Once in, I see two large, white furry faces peering through the spiked window.

Later, Olga, Lizza, and I continue our pursuit of data while we are lugged in the wooden box by the Buran, climbing hills and measuring muskoxen head sizes at close distances. My eyes have grown puffy and my fingers are cracked and swollen from the cold. Some days, we're beaten back by sudden whiteouts. On other days, we don't even make an attempt, and it's not always because of the Buran. When there are ground blizzards, we simply can't see.

Today, both weather and Buran cooperate. We hike more mountains and drifts and cross bergs of ice. We check our camera traps. White bruins are the most regular customers, some male, some female, and some chewing on the carcasses of muskoxen. Three of the bodies we check have bones with good marrow, a sign that the muskoxen did not starve to death. There are photos of bears with cubs and bears without them. One day, after the Buran fails, we still manage to gather data on muskoxen, but then posthole, sinking into the snow again and again, for an unarmed, stressful, six-mile trek to safety. Our database has grown to forty-two measures of young muskoxen. It's time to shift from photos to analyses.

To do so, we'll snowmobile thirty miles from Somnitelnaya to Ushakovskoye, a nearly abandoned village with the old meteorological station. It has a constant supply of electricity, and neither asbestos nor lead precipitate from the ceiling.

Some four hundred miles away, along the Alaska side of the Chukchi, biologist and coworker Blake Lowry and I had earlier come up with an ambitious plan: to attempt to simultaneously gather data on muskoxen along the shores of both continents. I'm keen to maintain my annual continuity of data in Alaska. The best time to do this is late winter. Since I'm already in Chukotka, Blake will lead on the eastern Beringia sites. He knows the data drills, having worked there previously with me. This way we can contrast body sizes and growth for populations on both sides of the Chukchi Sea.

Back in Ushakovskoye, Lizza and I enter the weather station by stepping down three feet of drifted snow. The chamber has four residents, none who've met an American. I'm "Jo-uhl" to them, a test.

Leanna, the full-time cook, is already happy. With Lizza, the island's female population swells to three.

Igor, too, is there. Ex-navy and current weight lifter, he's the station's mechanic. Most days he's tucked away in his tiny parlor, watching TV. In kindness, he insists I join him. We communicate by dictionary and hand gestures. His brother is still in Ukraine. Igor worries. The weather station has no internet or phone. He calls his brother on my satellite phone.

In a land roiling in white, a green and breathing plant adorns Igor's small windowsill. I admire the pride he takes in his meticulously watered specimen. His favorite possession is a different color—black: seventy-year-old vintage binoculars. One lens is corroded. The other carries the appearance of a fogged periscope in a *Red October* movie. How he sees polar bears and wolves with those optics remains a mystery.

With Lizza and me, Ushakovskoye's eerie human colony climbs to thirteen current residents. The island's three pet dogs double as bear deterrents.

Once a fishing and mining village that housed a small gulag, Ushakovskoye now consists of abandoned buildings that sit idle among concrete rubble. Drifting snow fills most of the structures, just as sand might fill a saloon in an old Western ghost town. Doors should swing open and close in a stiff breeze, but they don't—because everything is frozen. There is an old school with broken windows, its paint peeled away, a government building with spiked metal grates, and a post office. The library is dormant, filled with books coated in snow. The inside temperature there registers −9°F.

Tires and rusted chains are generously distributed throughout the town. Where the ground is swept clean of snow by the wind, shards of broken glass and plates appear. Thousands of corroded barrels leach oil, gas, and countless other toxics. Oxidized pipes and fuel tanks find camaraderie with shipping crates and bins. Primordial satellite dishes on rooftops creak. Battered boats that will not float, overland trucks that will not move, and engine parts that won't ever work again add macabre ambiance. Accretions of all that has been left behind cover the land from the water's iced edges to the nearby hills.

Amid this coastal devastation, the land has returned to nature. Arctic foxes move freely, their only worries being dogs and the next meal. Polar bears walk up and down what was once the main street. Their heads poke through windows or whatever else they wish. Muskoxen rub against the cemetery stones. This wintry Armageddon holds a surprise of a different sort from the freely roaming animals, hinted at by what we can now see from Igor's window: smoke puffing from chimneys.

Remarkably, seven souls call Ushakovskoye home, among them Olga. Another of the seven, Petrovich, whose first name is also Igor, has lived here for more than twenty-five years. All seven work for the Wrangel Island Reserve, the World Heritage Site, scavenging wood and molding parts from forlorn equipment to heat the very draft huts that serve as their homes. On this island, *Mad Max* characters have real faces. They use generators to produce their electricity for a few hours each day because they cannot use weather station facilities. In one of the tumbledown buildings is the equivalent of dial-up internet, but it rarely works.

There is no running water; the toilets are—hmm. The hearty Ushakovskians own not a single modern amenity. Whereas Alaskans split wood with axes, these Russians use a meat cleaver; instead of a light-weight avalanche shovel, they have the cast-iron version; instead of a vacuum cleaner, they fashion old grass into a broom. For rolling baking dough, they empty another vodka bottle. They are creative and generous survivors, with their government's blessing. Best of all, they invite me for meals, each of them regularly. These tough workers make but a few mere rubles and still they share their bounty, embracing me.

From the ruins of Ushakovskoye, a human existence rises. The trammeled patches here represent 1 percent of the island. Beyond, nature is raw and wild, its Pleistocene heritage yet palpable.

One of the heaviest polar bears northwest of our study area beyond Kotzebue, Alaska, weighed just over 1,350 pounds, which is almost the combined mass of three adult female muskoxen. Male polar bears are scary. They're also willing to tangle with any prey.

I wonder if muskoxen think about polar bears. Already our Alaskan work has revealed that they fear brown bears. Virtually nothing is known about interactions between white bears and muskoxen. Might a new predator-prey dynamic be emerging as a consequence of climate change? With more bears stranded on land, perhaps there are more interactions? From Hudson's Bay, we know bears are eating more berries, more grass, and more goose eggs. They also chase caribou. But how would anyone know if interactions with muskoxen are increasing unless someone had been studying them for many winters? The past literature offers little guidance on this matter.

"In the stomach of a bear which was killed in Scoresby Sound [Greenland] I found pieces of meat and skin of musk-ox, and not far away the carcass of the beast on which it feasted." So wrote A. Pedersen in the 1950s. What he didn't know was whether the muskox died through predation or from some other cause. On Kuhn Island, a trapper told Pedersen that when he found bear and muskox tracks so intermingled, he interpreted it as a predation attempt, though from what he could tell, there was no terminal ending and the tracks of the two species led in different directions.

That wasn't the case in 1932, when the following was written from elsewhere in Greenland: "Jensen found a musk-ox which had shortly before been killed by a bear. He had himself seen the bear while making the round of his traps. On his return he followed the tracks of the beast and came across a dead ox, still warm and showing marks which proved how it had been killed. All around the snow was trampled and bespattered with blood. With its claws the bear had badly lacerated the head of its victim."

Gashes do indeed offer important clues when not obscured by thick fur. My ears perk up when, back in Ushakovskoye, Ilya Borisovich tells about a male muskox with deep striations across its back. Ilya and I had interacted indirectly through common friends in 1996, when I spent some time with moose in Russia's eastern Primosrky Krai. Now, he tells me about the muskox bull that had been hanging around Ushakovskoye. One day the male showed up lacerated and lame. As a courtesy, he was shot. His loins were not wasted.

Elsewhere, polar bears do kill ungulates. On Svalbard, the Arctic Archipelago where the Pomors eked out their meager lives by eating reindeer, polar bears had preyed on at least five of these animals.

Interactions come with different signatures. Bears and muskoxen might exchange glances with, approach, flee from, or ignore each other. There can be charges, grouping, or passive expressions. The three muskoxen carcasses we had found earlier, all with good bone marrow, are suggestive of an intriguing dynamic—predation.

I pressed the issue, first with Ilya Borisovich, then with Petrovich, and then with another delight—a long-haired reindeer herder named Kaurgin, who is Chukchi. In 1962, the government mandated that names conform to "traditional" Russian ones. Little eight-year-old Kaurgin became the Gregoire Nikolaevich who I know as Grisha. Though related by heritage to Fred Goodhope on the other side of Beringia, much as my common heritage renders me a relative of Vladimir Putin, Grisha and Fred both having herded reindeer, but Grisha's roots to Wrangel are deeper than merely as a herder of reindeer.

Grisha was there when the twenty muskoxen arrived from Alaska in 1975. All of the muskoxen were less than two years of age and included seventeen yearlings—essentially babes on new tundra. That first summer, three were killed by polar bears: one, a female, had malformed feet and another had lung lesions. I exclude both from my sample of victims of predation, knowing the fencing designed to protect them was flimsy. Nor was there a herd with adults for protection. The third of the three newcomers killed was obviously also unfamiliar with Wrangel. Grisha is not unfamiliar with the island, having lived there with thousands of reindeer and having navigated it without GPS for thirty years.

I try to synthesize these observations to understand whether Wrangel animals deal with polar bears the same way Alaskan muskoxen respond to brown bears. Perhaps there is regional variation, or maybe the responses go deeper in history, from when mammoths roamed here and Pleistocene lions preyed on bison and probably muskoxen and more.

Wherever muskoxen are present, females form groups. Male bands are smaller. Polar bears often ignore bulls, but the opposite is the case when they encounter groups that are primarily made up of females. One time, two polar bears approached a female group that also included two mature males. The group fled. The bears chased them but were unsuccessful in catching them. Those same polar bears then approached a group of three males who remained steadfast, and the bears veered. Another time, a single polar bear lying in wait surprised a mixed herd that bolted in lieu of forming a defensive huddle. A female straggler was killed. Five events in all involved bears approaching groups of three or fewer male muskoxen; none fled and the bears moved off. Three other times, bulls and bears seemed to ignore each other. There were also four observations of bears approaching groups of females that also contained males, and the groups ran off. And twice, I was able to decipher interactions from tracks. In one of those instances, it was clear that a group with just a few individuals did not run. The other time, the group was larger and ran.

While the observations may be absorbing, the sexes were imprecisely noted, measures were not taken, and conditions such as snow hardness went unrecorded. To develop more insight on how different factors shape outcomes and perhaps the degree of recognition between species, I'll need to do playback experiments using visual models like I did in Alaska. I name my shamanistic reindeer Yakuts, in honor of the indigenous herders whose existence is reindeer. For the fake polar bear, I go with Eisbjorn, the Norwegian moniker for ice (*eis*) and bear (*bjorn*), something rolling off my tongue more easily than its name in Russian, *belyi* (white) *medved* (bear). I aim to do the playbacks when I next return to Wrangel.

As Lizza and I prep to depart the island, the crew at the weather station throws a party for us. I must have passed the Americanski test. Leanna wipes away a tear, and Sergey has imbibed more than normal. Koysta, the ex-rock musician from Moscow, delivers a bear hug. As sleds are loaded the next morning, I slip quietly into Igor's parlor and remove his vintage binoculars from the sterile windowsill. Next to his green plant, I place a new pair of Pentax. I depart unseen, smiling.

Getting home proves tough. On the helicopter ride to Pevek, it's only Lizza, a drunk, and me. He tries to assail her, she punches him in the face, and I intervene. The jet ride to Moscow is even more exciting. Two vodka-laden passengers brawl in the aisle until others halt the fisticuffs. The pugilists are duct taped to their seats. The remaining journey home lacks such passion but from start to finish requires a remarkable eighteen days. I use the time to sum up the photogrammetry findings from Chukotka's stunning island.

Wrangel juveniles were smaller than those from Alaska in every single year. That is, the one-year-olds, two-year-olds, and three-year-olds of both sexes were smaller than Alaskan counterparts across each of the eight years of measures. The probability that such size limitations would occur by chance is less than one in a thousand. The differences are not attributable to variation in genetic origin since all Wrangel and Alaskan animals are descended from the captures in Greenland in the 1930s and, ultimately, from Nunivak Island animals. Most likely, the short summer growing season and more frequent rain-on-snow events on Wrangel were responsible. While it's not surprising that weather affects body size, what's of interest is how nuanced and powerful the effects are. We found that rain-on-snow events that occurred during muskox pregnancy retarded individual growth for at least three years.

Olga and Sasha promise I'll be invited to return.

After my reappearance in Montana during 2014, I take a sabbatical and live on an alpaca farm in Colorado. The next year I'm offered the Barbara Cox University Chair in wildlife conservation at Colorado State University (CSU). Despite my love for Montana, I accept and prepare for a state more crowded and with more sun. As a bonus, WCS coordinates my continued affiliation with them but at CSU.

The year 2014 is fraught with tension between Russia and the Western world, which expresses its dissatisfaction with Russia, especially after the mistaken yet deliberate downing of a commercial Malaysian jet in eastern Ukraine. Meanwhile, Russia reconstitutes its military strength and constructs a new base on Wrangel in violation of the UNESCO agreements they had signed a decade earlier.

The Wildlife Conservation Society—which operates in sixty countries and has more than 350 field projects—is intrigued by the Russian actions and deliberates about its own position at an upcoming protected area convention in Germany. I wonder what the consequences will be if we sign onto an avowal opposing militarization at World Heritage Sites, given our existing collaborations with Russian coworkers on Wrangel. Beringia Program director, Martin Robards, along with Dale Miquelle who heads the Amur tiger program, agree that the cooperative work is what's most important. In Germany, WCS offers no commentary about the new base on Wrangel.

Timing becomes too tight for me to do fieldwork in Russia in 2015, so I head to Alaska and work there instead, this time with Gana Wingard. She helps deliver talks in six native villages, a bonus for all. Who could be better for this than a Mongolian conservation biologist? Gana's ancestry renders her more palatable locally than me.

Sasha follows up, indicating that despite the military presence on Wrangel, I can come back in 2016. I'm viewed as an asset and have caused no trouble. With a study plan in place, all looks promising, and Russia's Ministry of Natural Resources in Moscow offers an official invitation. I then wait weeks for my visa, but it never arrives. I pester Sasha in Moscow and the Russian embassies in DC, San Francisco, and Seattle.

There are now only seventy-two hours before I'm to fly, and then forty-eight. I'm uneasy about yet another tightrope, but finally I've got the visa in hand. Then, somehow the SIM card for my satellite phone was sent to the wrong address. This is not good because it's my safety net for an Arctic emergency. Morning dawns, and the shuttle to Denver's airport should arrive in an hour. Before seven, there's a pounding on my door. It's a delivery person with the SIM card.

The *US News and World Report* runs an op-ed I wrote on geo-diplomacy, touting how an unlikely furry creature of the Arctic can find common ground in and be of benefit to mutual Russian and American interests. Soon thereafter, the U.S. Embassy in Moscow asks me to come in. I feel tentative about it but agree. I also consent to giving a public lecture. Eighty Muscovites attend. Most of the follow-up questions concern what ordinary Russians can do to combat global warming. Some ask about the consequences of climate change for wildlife. The attendees are diffident but draw courage and pose queries in English. Being unfamiliar with Russia's media and popular dissent, I'm cautious about what I say, knowing only the hopelessness frequently expressed about Russians by Western media.

I tell them to talk to their friends. I suggest that if every person can get four others to speak out and write to the government officials about energy use, and these in turn do the same with another four, and four more follow, pretty soon a third of a million voices can register through social media or real letters a stance on climate issues.

I then head far to the east. Lizza is not with me this time. Instead, Sergey Abarok will be my interpreter. The son of a sailor from Vladivostok, tattooed from shoulder to wrist, motorcycle rider and musician, he's spent time in Southeast Asia and once visited New York. He now runs a Russian ecotour company. I'm slightly concerned because he has no Arctic experience, but beggars can't be choosy. He tells me not to worry: "I'm Russian, I can handle it." We join forces, and arrive together in Pevek.

Immediately, I'm detained.

Before my court hearing, I receive an email from my daughter: "Dad, this sounds serious. I'm worried. Are you OK? Please call or email me after you appear so I know. I love you, Sonja."

Sergey is only admonished. Not so the Americanski: I am found guilty of illegally entering a restricted military zone (the province of Chukotka). I'm fined. But this isn't over. If Moscow doesn't offer additional clearance, my venture to Wrangel will just be a time-consuming diversion. The courts will be less kind once I'm off the island. At least the gulag at Ushakovskoye is closed.

Sasha maintains it's not a big deal and that the heads of border security in Moscow are just slow. Hmm, easy for him to say: he's not been interned in the United States for illegal entry. After I insist, the Wrangel Reserve covers my fine. I'm delighted that WCS is more concerned than Sasha and considers the next steps. Dale Miquelle doesn't want this to escalate, and neither do I. Together we elect not to inform the U.S. Embassy in Moscow. I'll lay low and be zen. Sergey, Sasha, and I head to Wrangel on an Mi-8.

Two surprises wait.

Shortly after we land, Sasha climbs back aboard the helicopter, shakes my hand, and bids me a good field season. I was saddened to learn from him that a different crew has replaced my surrogate family at the weather station. In a swirl of snow and blue haze, he's gone. The second to go is Olga. She had clashed with the master of communication efficiency. With no one to replace her, Sergey will be my field hand.

The military base commander on Wrangel is, I'm told, a true patriot and upset that my visa was approved. I wonder whether he's had anything to do with my detainment in Pevek. Maybe he's paranoid. Perhaps it's only me who is paranoid.

The Wrangel fieldwork is basically a repeat performance, but with a few sidebars. Muskoxen are widespread, sometimes locally dense and, sometimes, miles and mountain ranges apart. I traverse more than 550 miles on ancient snow machines including the Buran, and spend three weeks in the island's remarkable and colder interior. Four of the five staffers get frostbite, having refused face masks for cold protection. Their proud refrain: "We're Russian." Sergey's toes grow numb and blister, and he misses fieldwork. My visual deceptions—the shamanistic Yakuts costume and its partner Eisbjorn—work well. Petrovich is a splendid and patient guide. He tells me when my toes go cold I can drive the snow machine rather than grow ice on the back seat or in a sledge. Frostbite is not my only problem.

One day when I'm getting data as Yakuts, a solitary polar bear, the real thing, appears. Male polar bears are especially bold. It's not as if the tactic I'd tried in the past—the airborne head as an anti-death mechanism when my costume was loosed into the air and I shifted quickly to Mr. Biped to stop the muskox charge—will thwart a polar bear. And one is now angling toward me. Besides, this is the Arctic. There is no retreat. Running and hiding are not options. My single recourse is hope that the bear realizes a reindeer, even a fake one, is faster than he.

The bear ambles slowly.

Two miniscule cubs follow. Whew. Not that females are not dangerous. In amber light and long shadows, the trio changes direction.

On another day, bear tracks lead to a cliff smothered in wind-whipped drifts. The snow is piled, and a den entrance dug. Armed with a fifteen-inch flame-spewing torch and a ten-foot stick, we investigate. As the distance to the den diminishes, my heart races, but no one is home.

Over the field season, there are more muskoxen, more wolves, and more polar bears. On one slope, a fresh set of wolverine tracks approach the fresh tracks of muskoxen, along with the deep prints of ice bear and light ones from a white fox. A different time, I watch a seal on ice a couple of hours before dark. At first light the next morning, a bear is at the identical spot, slamming massive paws under the full weight of her body trying to get to the seal in an ice den. One smashing attempt follows another, and another, and another. Ninety minutes later the bear departs hungry. Two newborn cubs trail.

After more than a dozen visual experiments with muskoxen, half as Yakuts, the mad shaman, and half as Eisbjorn, we prep to head to Ushakovskoye to find more muskoxen. With three federal entities now on Wrangel—military, meteorological, and nature reserve—I expect political shenanigans. There are none. But a note scribbled in Russian is left for me in a cabin. It's by the commandant of the military base. It simply says hello to the "Americanski."

Days later, four troops arrive in a prehistoric vehicle to drop off thirty solar cells and a trailer filled with wood. Dimitri, the forty-something commander, says its fine for me to ride in their Vezdekhod tank—an over-snow primordial beast with a turret-like chamber—for the thirty-mile ride beyond Ushakovskoye but commands: "No photos." After a lunch that includes vodka— Dimitri's lunch, Petrovich's vodka—Dimitri removes his uniformed jacket and asks for a picture with me. My unkempt and unwashed hair is the perfect antidote to his finely coiffed head, manicured military style. He later points to a mirage of floating icebergs: "New York, after nuclear." He does not smile.

I descend into the intestines of the Vezdekhod and enter a gaseous fusion. One of the soldiers already there, a tooth missing and another capped in gold, lights a cigarette. *Really? OMG.* Gears, valves, and pistons grind. I don't know which is louder, the roar of an Mi-8 or the pounding in this doleful chamber. The windowless ride is long. Intoxicated by heavy fumes and deafened by engine noise, I stumble out, back into Ushakovskoye's devastation for the first time in two years.

During the ride, I dreamed I was back at the weather station and that the personnel had not been replaced, that I receive Russian bear hugs and kisses from all, that Igor has the same green plant in the window and treasures his new Pentax binoculars, that Grisha's hair remains uncut, that Koysta still rocks to music, and that Leanna still cooks. It's a *Wizard of Oz* moment: Dorothy wishing to be back in Kansas.

Sometimes, dreams actually come true.

As I step back onto Ushakovskoye's soil, I'm showered with smiles and meals: Sasha had been mistaken. To my surprise, the crew I had known from the meteorological station were all still there.

But my temporary residence this year in Ushakovskoye is four hundred yards away from the meteorological station, with a very masculine team. There is Sanya, a Russian ex–Navy SEAL equivalent; Petrovich; and several compassionate inspectors, what Americans would call rangers. The quarters are tight, with four to the single room that makes up the entire decaying facility, complete with rusted drums for water, greasy and paint-scaled walls, and again loose lead and asbestos particles as a bonus. They tolerate my early morning coffee and my rehydrated meals and appreciate my efforts to work communally. They grow fond of my Americanisms.

The field season is cold. On one day with wind chill down to −47°F, I just cannot warm up. My fingers lose feeling and then burn painfully as blood finally tingles forth again. I jot a note.

April 5, 2016

Hi Sonja,

I'm writing because today was a bad one and this helps. So selfish, I know. Fieldwork today was miserable. Sometimes I wonder why when your friend's parents are retiring, that I don't. Maybe this will help me figure that out.

I was hoping to get warm after bouncing on the snowmobile. I sit on the back and get progressively more like a piece of meat frozen hard in a freezer. OK, so, slightly embellished. But, when I finally re-entered the cabin, I managed to feel toes and fingers. It's not great. And, everyone smokes all the time. The haze and the smell is—as you can expect—disgusto.

There is no place to avoid it. My throat burns. None of the Russians think twice about lighting up. Second hand smoke—what's that? I remember a Jimmy Hendrix song 'Purple Haze,' but truthfully I have no clue what that song is about. Once I have Internet, I'll look it up. But, there are good days too.

I've been able to watch cuddly little muskoxen with their mothers in sun when it's 8 degrees, and it felt so warm that I wanted to trade winter clothing for short sleeves. The last few nights the sky has been brilliant under the fiery aurora. There was a wolverine feeding on a dead muskox, and a few days ago I watched a female wolf mate with a dog among the rubble of an abandoned village. We've seen bear tracks among pressure ridges, and newly born cubs out from their den. Dimitri—the "cap-EEE-tan" of the military base—sent a loaf of warm bread for me before we all disappeared into the island's interior. A fellow named Petrovich gave me a Russian sniper suit so I can become one with the snow for my playbacks. The Navy seal guy I told you about handed me a rocket-launching gun for protection. The land really is ominous in silence.

And there is this from a few days ago.

One group of muskoxen had been really troublesome. They're about 15 miles away, which means it takes more than an hour to reach them by snowmachine. And, so a few days ago, we were getting ready to do playbacks, but the wind changed direction and they heard us. My fault. So, two days later we tried again, and did the fake caribou. Then, yesterday, freezing fog drifted in from the coast where the pack ice had been split apart by wind. So, we went out to look for the same muskoxen because I still needed to do the bear experimental approaches. But, because of the fog we couldn't find them. Finally, yesterday, we succeeded. Honey—this is probably boring you, but here's the point.

Three days, almost 100 miles of riding from 5 above to 20 below, just to get a single data point. One! Just crazy? And, by the time I submit our work for publication—well, who knows, some reviewer will probably say, 'sample size is too small.'

OK, that's it from my frozen box and my uplift for today. I'll fill in more in a few days.

But, I know why I do this. The land and the animals are special. The people too.

More about you in a bit.

Love you.
Dad

Of Beringia's large mammals, twice as many live in the sea as on land. Polar bears are dual citizens, in that their very existence is coupled to ice yet they can still spend months on land. Brown bears differ on this score, of course, but, like polar bears, are also being squeezed. They can't go south—too many people. They can't north because the land ends and they don't live on ice.

Ecological systems are colliding with bewildering regularity, and when species move, novel encounters occur. Brown and white bears now encounter each other, as well as muskoxen. We don't know if the interactions are more or less than in the past, or the same. Only now are we learning about outcomes.

On Wrangel, the muskoxen respond to white bears about the same as their ancestral stock react to brown bears in Alaska: with vigilance. There are differences, though, too. Polar bears sidle closer to muskox groups than do browns in winter, which might simply be because they blend better: white on white rather than brown on white. This situation is not quite the climate-induced color mismatch of hares, and, in fact, in summer, it's possible it may be the opposite, with browns getting closer.

But here's the relevant point: even though white bears don't seem to be better predators, muskoxen are flightier when they interact with white bear models than are muskoxen on the Alaskan side when they see fake brown bears. Irrespective of locale, female muskoxen are more sensitive than males and thus are less likely to stay, perhaps because of size (smaller animals are more likely to be eaten) or maybe it's a maternal thing (females invest more in offspring than do males). Either way, males stand their ground more often than do females. Nevertheless, both sexes react to bears as if they are serious predators and differentiate fake caribou from fake bears—and flee more often from white than brown bears. The difference in likelihood of fleeing is not because of snow conditions or group sizes, because these are factors controlled in statistical models. Maybe the response difference arises because, while brown bears handle heat better and so can endure long chases, polar bears are not good at long pursuits due to overheating, likely in part, because, they are more aquatic than terrestrial. Perhaps muskoxen have learned that, in the case of white bears, running offers better dividends than standing defensively. Maybe standing against white bears is just deadly, so flight might be the best of bad options. The answers, at this point, are more uncertain than certain.

Another difference between eastern and western Beringia is that muskoxen are free from harvest on Wrangel, and so their sex ratios are not skewed as in Alaska. Offspring survival seems higher on Wrangel. Still, polar bears are bears, and bears learn quickly about new sources of food—whether potentially easy prey or something else.

On Wrangel, that something else is trash: bears and polar foxes have figured out that, because the military on Wrangel has not effectively enforced trash disposal rules established by the biosphere reserve, they have easy access to food.

When up to eighteen white foxes descend at a time and fight over the rights to lick the grease from a fish tin, or when polar bears visit the garbage bins, becoming habituated to such easily won meals, or when, on a dare, a soldier feeds a bear by hand, further habituating them to close contact with humans—internecine anguish can result. A mischievous contractor or two bait food with an explosive, then watch as the bear's face ignites and as the poor bruin writhes uncontrollably in agony.

How do I know this? I've watched secret video clips handed to me discretely by someone on the island who was wishing I could embarrass the Russian government once I'm back in America.

But I like the postapocalyptic island, and I wish to return, so I've made sure the files are now gone.

When John Muir visited Wrangel in 1881, there were neither wolves nor wolverines, reindeer, or muskoxen. Three decades later, when Bartlett's expedition found themselves on the island, they found none of these either.

In 1948, settlers introduced reindeer to Wrangel. Subsequently, Grisha herded them, danced with them, and dressed in their fur. It was 1975 when the muskoxen arrived from the United States. Three times since then, wolves discovered the island. Wrangel is far from terra firma—eighty miles at its closest point—and reaching it is a gamble, a bit like being an astronaut spinning wildly through space and hoping to arrive on a planet with oxygen, food, and shelter. But wolves are good colonizers. They got to Isle Royale in Michigan on their own, and those that made it to Wrangel navigated twice that icy distance. The Wrangel wolves have gone extinct at least twice and are still persecuted, even after the reserve was established. Wolves, Wrangel Islanders have told me, are not a good species like muskoxen or the now-feral reindeer. But it may be icing and rain-on-snow events, and not wolves, underlying the reindeer collapse from six thousand to fewer than several hundred today.

Wolverines have also colonized Wrangel, managing to cross leads and labyrinths of ice and pressure ridges to get there. Wrangel is but a mere dot in a vast Arctic Ocean. So not only did one wolverine have to be lucky but so did a second wolverine—unlike with luckless George, the Pinta Island tortoise who died a lonely mate-less death. They had to find each and procreate, which they've done. Although it's also possible that it was only a pregnant female who migrated and later gave birth. Either way, how they subsist is uncertain, though we do know they eat carcasses: I've watched two feeding jointly on a muskox. Maybe they also eat reindeer.

This is a place rich in history, one of obliteration featuring empty buildings and decadent piles of tires, rusted boats, and hulks of deadened machinery and a few splendid souls. This is also a land of transformation, beauty, and remarkable nature to which an eclectic array of species have been imported or have colonized on their own. Among the latter group, as I've said, are the wolves, though they are not really wolves, some being all gray, others black, some having shortened, dog faces, and others having more "normal" wolf snouts. And so they are persecuted as hybrids, the reasons for which I will address shortly. Can one really tell a wolf phenotype at a glance? Probably, but only when the distinction is extreme.

Which part of a genetic continuum do we call natural? Think bison. Think yak. And what about us? We house remnant DNA from Neanderthals. Are we hybrids?

As the Arctic warms, species from the south increase their northerly presence. We see this increase among parasites and the diseases they carry, among robins and ravens, and among moose. We see it with bears in terms of their interactions with each other. On Barter Island, in the village of Kaktovik in eastern Beringia, for instance, grizzly bears dominate polar bears at bowhead carcasses.

With the advent of climate change, there is an increase in the overlap of terrain for the two species, and mating between them has been known to occur. The hybrids, known as grolars or pizzlies, were thought to be confined to zoos, but more recently they've been found in Canada's Arctic, where grizzly bears are increasingly sighted on islands. The result is some overt phenotypic mix—faces, ears, snouts, humps on the shoulders—reflective of a bear of the white variety or brown, which we know because the few unlucky enough to have been shot gift us their DNA, which provides clues to their hybridized past. The number of known wild hybrids now approaches ten. At least once before in the Holocene history of these two species—the seal eater and the omnivore—introgression (or hybridization) took place: DNA from bear remains found in southeastern Alaska, from when glaciers were deeper and more widespread, confirms this. Climate—it can unite and it can disassociate land and its species. I think hybrids and contemplate bears. Others consider a different suite of carnivores.

So, to return to the wolf, is the wolf phenotype really so obvious? Wolves carry dog DNA, which is sometimes coded for black coloration. Yellowstone wolves come dressed in different morphs, including some that are black. Black, in short, is a heritable trait derived from introgression with domestic dogs.

Why on Wrangel, or anywhere else, are wolves or wolflike dogs being shot? Easy: either they are not tolerated because they are a "bad" species, which kills other animals, or they are just killed for sport. The same is true for grolars and pizzlies—it's not for food that they're killed. The answer lies in the human psyche and our preferences for what we think nature should or should not be.

Among the people who have considered this is a bright and daring woman named Emma Marris. Her provocative book, *The Rambunctious Garden*, has created quite the storm. At one of her lectures, which I attended with a few hundred others, she suggested that we cannot turn back the clock, that the introduction of invasive species and other changes are natural. She's right, depending on how we define "natural." She argues that there is much we can't control and that we should just let nature play out, with us as designers in chief.

Far earlier, in 1859, Charles Darwin touted such oft unappreciated blending of wild and domestic in his *Origin of the Species*:

> Humble-bees [bumblebees] alone visit the common red clover (Trifolium pratense), as other bees cannot reach the nectar. Hence I have very little doubt, that if the whole genus of humble-bees became extinct or very rare in England, the heartsease and red clover would become very rare, or wholly disappear. The number of humble-bees in any district depends in a great degree

> on the number of field-mice, which destroy their combs and nests. . . . Now the number of mice is largely dependent, as every one knows, on the number of cats. . . . Hence it is quite credible that the presence of a feline animal in large numbers in a district might determine, through the intervention first of mice and then of bees, the frequency of certain flowers in a district!

I raised my hand, following Emma's lecture, and asked if she had seen any of the *Mad Max* movies. When she confirmed that she had, I asked her if that's the world we want, one dominated by feral cats and dogs, one where wildlife is vaporized—basically, a world run amuck that we accept calmly.

Fortunately, that apocalyptic vision does not apply to Wrangel Island. Apocalyptic places do exist, though. History has taken care of that, from radioactive atolls to Chernobyl and beyond. But even such places as these can and do have a future. At Chernobyl, there are now moose and wolves, wild boars and beavers, though the number of people there remains few. Although the land is being reclaimed, it is different than its baseline.

In short, we just need to know what we want, agree on it, and determine how to get there. Which is the more onerous challenge: climate-altered environments? Or the ones whose landscape changes we quicken in our palms? A binary world it's not, though one hand massages the other.

Imagine no possessions. I wonder if you can.
No need for greed or hunger. A brotherhood of man.
Imagine all the people sharing all the world.

JOHN LENNON, "Imagine," 1971

17

Nyima

Suppose you're going on a mission somewhere up high on earth's surface. The purpose is scientific discovery. You'll be remotely placed, and the conditions promise to be primitive. You hire three cooks. Their names are Sunshine, Hope, and Darkness. You're not in charge of food planning, but I am. I've done poorly. Supplies run low.

Sunshine says to the team, *Take mine.*

Hope says, *No—mine.*

Darkness says, *OK—thanks—I'll survive.*

In Tibetan, *nyima* means sun, *tsering* hope, and *khog nag* dark mind, often selfish. *Nyima* and *Tsering* are the names of real cooks who accompanied many expeditions in northwestern China.

In Mongolia, Tibet, and the Arctic, living in groups improves survival, much as it does in many parts of the world. My life for fifteen years has involved living and traveling in bands—teams reliant on each other and united by common cause, independent of ethnicity or culture. It has required getting along, cooperating, and seeing the world through the eyes of others. Life, in short, is more than science.

We're a social species. The word "social" connotes different things in different contexts, but there is no denying a link between society and civilization whatever the frame of reference. Some constructs are good for society—there are those who would argue that socialism is not among them. But socialism spans a continuum of ideologies, and, like the words "society" or "social," it carries varied meanings—"communal," "community," "collective," "communism," "Leninism," "Maoism," and "Marxism," among them. We're all living in societies with geographical and political boundaries, but each of us is also a global citizen.

Many societies are principally democratic and accompanied by individual freedoms, yet even these types of societies are not necessarily just. Eminent domain punishes individuals while arguably achieving a more virtuous goal. Women and minorities have not always been treated equitably. And while surging capitalism has brightened the future for millions if not billions, generating both dollars and ingenuity, it can also create brutal asymmetries. While some individuals struggle to merely exist, others achieve the stature of a Donald Trump, building and heating twenty thousand–square-foot homes or acquiring a fleet of Humvees—and our legal system protects such immense beauties, regardless of their carbon footprints.

Profligate spending, disparities in the distribution of resources, and other economic forces result in both societal challenge and opportunity. We, along with animals and plants, compete for air, water, and food. We're all in this world together.

When our resources run perilously short, who do we wish to have on our team—*Nyima*, *Tsering*, or *Khog Nag*? More decisively, do we make that wish a conservation reality?

Alexander Gruzdev once asked why I travel twenty-one times zones to a field site and try to do conservation and science when many have given up. Despair is evident in the faces of everyone from my Wrangel Island comrades Petrovich and Borisovich to ordinary Russians. People shuffle with the indifference characteristic of the oppressed, but Sasha does not. He smiles and builds tourist facilities on Wrangel in anticipation of establishing a place of education and communication and science matters. Similarly, the eighty attendees at my lecture in Moscow wanted to know how to make a difference, and John Goodrich and Dale Miquelle have toiled for decades with local teams, now using the data they collected and every means possible to sustain Amur tigers. Clearly, not everyone has grown indifferent.

Neither has Tshewang Wangchuk given up: he continues to press for environmental awareness and education to better Bhutan's biodiversity. Tempa Tshering leads the charge to protect their largest cat, the tiger. Aili Kang left the sanctified air of Lhasa for starless Beijing, and then went on to the Bronx, where she now directs WCS's Asia program. Buuveibaatar represents a new generation trying to bolster conservation in the Gobi, as does Ganchimeg. Ellen Cheng departed America's comfort for the privilege of sleeping on floors in Mongolia and Bhutan so she could tutor students in conservation science. Baima guards against poachers in Tibet, as does Zhao Xin Lu in Kekexili. Freddy Goodhope shoots caribou to feed Shishmaref's elders.

Ubiquitous are the people with the names Nyima and Tsering. They work to better the lives of many species, small and unknown, big and sexy. They ask little in return. They labor in their backyards or far afield. Some work for NGOs or governments, some for schools and universities, as teachers, electricians, and educators. Others participate in recreational activities, such as and hunting, fishing, and hiking. Many are students, gardeners, or just observers of nature. They come in all flavors—from educated to uneducated, old to young, Nicaragua to Nigeria.

When poor Rwandans living among the highest human densities on earth can hand mountain gorillas a future, and Namibia's desert rhinos can be conserved free from fences, should we not expect successes of similar magnitude in high and in cold realms? It's not just human density that matters. Governance, lifestyles, and choices do too.

I've concentrated largely on cold-adapted species of the snow oxen ilk—how they got where they are, whether and when they adapt, and their prospects for continued existence. I've asked if their Pleistocene ancestry imprisons their future or whether they can keep apace of our reformatting of the planet.

Neither are zero-sum games—nor is climate forcing or habitat alteration. Shades of gray blanket Earth, as opposed to the black-and-white binary option of life or extinction. There are those who even question the concept of extinction because the DNA of dinosaurs, mammoths, and passenger pigeons persist in contemporary forms. Given the level of sophistication of our science, the possibility of resurrection is not out of the question. But what happens when we've run out of places to put the genomically engineered? Changes to the world result in less habitat for plant and wildlife, while at the same time producing more people and species that depend on them, more ecosystem novelties, and new hybrids.

Most people realize we persist in a singular world connected by more than the internet. Nature and human ecosystems are so strongly linked that it is no longer clear where one stops and the other starts, causing challenges to intensify.

Fortunately, some people also realize that the choices we make shape how and where species live and for how long. Despite the horrifying regularity with which we eradicate habitats and release carbon, lifestyles are controllable. Options open up.

Our reliance on fossil fuels is changing, and alternative energy sources are increasingly available. Multilateral agreements about wildlife and other trade, about energy, and about climate are signed at global scales, hopefully with some teeth. While global megacorporations generally oppose reigning in carbon emissions at every step, gains still come. The use of lead in gasoline was banned in the United States in 2009; other countries followed this lead in pursuit of cleaner automobiles and air. Europe has undergone greening, sometimes at a faster rate than the United States. Americans listen and watch, notwithstanding ignorant politicians with an antiscience bent or red herring claims of too much socialism. Yet the planet's carbon mat will be lessened only sluggishly because we can't quickly redirect our titanic love affair with combustibles.

We can, though, manage our inevitable demographic growth and our zest to populate all the planet's habitable corners. We can disengage the propensity to propagate our genes from Darwinian prophecy and limit our reproductive rate. Increased access to education and improvements in economic development lead to lower birth rates and gains in equality for women. Restraint can be a virtue in that reduced human reproduction will enhance opportunities for the planet's biodiversity.

Conservation successes have been achieved by aggressive trials. The protection and maintenance of open space for wild yaks, chirus, and kiangs have netted them population gains. Under watchful eyes, saigas have not been slaughtered for their horns, but, unfortunately, have nevertheless been challenged by *zud* and mysterious diseases. Muskoxen have been safeguarded, restored, and reintroduced into several habitats, but rain-on-snow events can kill thousands in a single setting. Countless animals of other species, though, still die in illicit trade, and rangers and environmental advocates on every continent are murdered. The business of conservation is sometimes stained red.

As sea and land temperatures warm, moisture regimes change everywhere. Ice will continue to melt at high latitudes and elevations, even while deepens in some places. Under these conditions, yaks and muskoxen both do poorly. The Seward Peninsula is modifying quickly, with its vegetation growing more profuse as its climate becomes increasingly maritime—conditions that render the future of its muskoxen less sunny than the colder and drier eastern Arctic or the Taymyr Peninsula of far northern Russia. But where population sizes are not terribly small or highly fragmented, extinction still seems unlikely. With sufficient space and human hunting under control, we should be able to avoid a zero-sum consequence in the future.

In the vastness of the north, caribou require big lands. They are on the run, what with lightning increasing, permafrost melting, lakes draining, and tundra fires becoming more frequent. But at least there are few people so there is little need for fencing, even if roads, energy infrastructure, and mining are not going away. Caribou don't notice the unsightliness of pockmarked places, well pads (areas cleared for drilling rigs), or thousands of rusting containers. They can live without scenic splendor. They migrate anyway and can persist if the permafrost is not melted out under them and rain-on-snow events moderate.

The highlands of Bhutan and Tibet offer remoteness and promise of a different type, the former having already created bio-corridors, and the latter having established a vast network of nearly connected lands—larger in size than California—that are protected. Such places offer wild species the opportunity to thrive. Opportunities to help animals also exist in more crowded and temperate zones. Even where fences cannot be undone, they can be made friendlier by removing rusted barbs and raising the lowest strands so that fawns and calves and the like can follow adults during migration season. Travel corridors can be protected. Under- or overpasses can even be built to connect habitat patches and sustain migration and gene flow; such structures are now found from Singapore to Germany, and from Tibet to Los Angeles.

Science has played a critical role in guiding conservation efforts by indicating the best locales for corridor creation and helping to curtail the spread of disease. Scientists have given insight into the lives of species, generated fascination in the process of adaptation, and revealed the relevance of ecosystem function. But progress will cease if we hesitate to identify where the causal problems lie because solutions aren't possible without first knowing that. The fusing of science and conservation reveals why most long-lived species, including tribal animals such as snow oxen, aren't likely to be able to adapt and thrive in our radically altered ecosystems. Their survival will only be possible if altruism exceeds greed.

Science has also helped improve the understanding of ecological baselines—a pursuit with about as much cachet as conservation biologists have in some parts of Alaska, which is to say zero. But such knowledge has nevertheless led to actionable conservation. People are surprised to learn that muskoxen went extinct in Alaska or that wood bison once lived in Alaska's boreal subarctic less than two centuries ago, as Georg Steller was exploring Russia's America. Ultimately, knowing this past led to a restructuring of the future and, through reintroduction, both species are back on home turf in Alaska.

South of the great north, other bright spots have appeared. New England has regained its forest cover, more than doubling the 30 percent of forested land that was there late in the nineteenth century to about 80 percent today. The presence of white-tailed deer there supports populations of canids, though a mix that includes coydogs and wolflike dogs. Moose are back in New England's forests, as are turkeys. Out west, Canada's Nahanni National Park Reserve was nearly tripled in size by First Nation consent when it became clear that its boundaries were too small to sustain wandering grizzly bears and caribou.

Less than thirty years ago, few would have imagined the number of large carnivores expanding in a country with more than three hundred million people, where once those animals were shot, poisoned or fumigated. Now Montana, Wyoming, and Idaho have more than fifteen hundred wolves, and almost two thousand grizzly bears roam the Yellowstone and the Crown of the Continent ecosystems. Black-footed ferrets, once extinct in the wild, are back on the ground in Arizona, the Dakotas, and even Mexico. Condors fly over Arizona and California once again, and the gigantic Bolson tortoise has been returned their native habitat in the Chihuahua Desert of New Mexico.

Alien predators have been removed from some eight hundred of the world's islands, improving the survival chances of petrels, boobies, and kiwis. Large parts of Europe have been rewilded. Brown bears and wolves are in portions of Italy, France, Spain, and Germany, and tigers populate parts of India.

Nyima and *Tsering* disclose guarded smiles.

We've handled the world with dirty gloves, with grave consequences. Perhaps 90 percent of the population do not know what's at stake, do not care, or feel too helpless to do anything. Other than the air, water, and food that have been affected, does it really affect their lives?

Conservation, rather than being seen as fundamental to the survival of all of us, is regularly denigrated by too many as just another "special interest" or as radical environmentalism. Can't this change, and can't we do better? Optimists say yes. Pessimists suggest no, our destructive path will continue and overwhelm us.

At the same time, an unerring appreciation for nature continues. More people visit American national parks in a given year than all professional sporting events combined. Zoos attract about the same number. People obviously care about nature and its inhabitants. Some eight billion enjoy some form of ecotravel annually. The link between environment and economy is strong.

Yet there is also an overarching view that biodiversity does not affect our lives. That may seem true for many, but nature manifests two fundamental ingredients that influence us greatly, agriculture and disease. Food security and health are foremost in our lives, yet connections involving biodiversity and our living world often fall from many on the public radar. Do we really need a grisly experiment to demonstrate our deathly séance with nature?

If so, that experiment might alight again at the intersection of wildlife and disease—bats disseminating the essence of the Marburg or Ebola viruses, a different zoonotic that afflicts migratory birds, novel viruses in fish, or mosquito-borne pathogens? Maybe it is only when massive numbers of "regular" people in Western nations die from some cataclysmic disease event—clear, undeniable, and interfaced with biology, nature, and climate—that boundless self-interests will no longer undermine the common good. And, maybe, that's when ignorant politicians will realize that it will take not only thinking ahead but also the real implementation of plans to assure environmental resiliency will produce true benefits for wildlife and for people.

I live just outside Fort Collins, a bustling university town snuggled at the base of the Colorado Rockies. Wide streets intersect with quaint ones, and cute homes and cookie-cutter mansions pervade the area. Commuters follow bike paths along waterways to their offices. Green spaces dot a cosmopolitan landscape; the favored mantra is a park for every square mile. The quality of life here is good despite soaring property values and painful decisions about growth.

Fort Collins wasn't always this way, and some resent the changes. Burning wood for heat is no longer allowed. Vehicles require pollution abatement. There is a rush hour.

The town and the county have purchased reasonably sized chunks of land, amounting to a few hundred acres within the city for parks. Farther outside city limits, they own several thousand acres more. On Fort Collins's public properties are rattlesnakes, prairie dogs, and burrowing owls. Also, reintroduced bison and black-footed ferrets are found on the more distant of this public property. There are coyotes and cougars.

Denver, which lies sixty miles down the road, has more serious congestion and air pollution than does Fort Collins, though both have thriving economies. Denver, too, has invested in parks and nature, sometimes in conjunction with the federal government, and an exciting yet controversial exurban wildlife reserve is set to open there: the five-thousand–acre Rocky Flats. The Flats, once storage facility for plutonium whose toxic residues led to a $7 billion renovation, was not the sort of place for kids or dogs. It's now a wildlife refuge, arguably cleansed. People have seen eared grebes and great blue herons there, as well as, occasionally, golden and bald eagles, and there are bobcats and elk. This is restoration. It is conservation. In many ways, it is the future, though it will be an expensive future.

Change occurs because people demand it. They continue to achieve positive change through communication and outreach, through instigating, lobbying, and educating, and finally through the ballot process. They achieve victories at scales of the municipality and the county, all the way up to state and national levels. As with Path of the Pronghorn, voices can deafen when groups join forces. Already conservation advocates wear many hats and have differing but connected goals; from the Jackson Hole Conservation Alliance to Ducks Unlimited, from the Wilderness Society to the Sierra Club. Others are more international, from the likes of WCS and the World Wildlife Fund to Conservational International and the Nature Conservancy. All do their part.

Scale matters, which in my opinion is the critical reason to protect big and remote areas. The world is no longer producing large bio-diverse regions, and by cannibalizing what remains, we squander opportunities to assure the space needed for some of this planet's most spectacular animals. But poor land use is not the norm everywhere, and alternative visions of resource management can lead to precedence and to hope.

The Kavango-Zambezi Transfrontier Conservation Area, or KAZA, is one such place. This massive designated conservation zone of 109 million acres has a relatively low human population. It spans thirty-six reserves across Angola, Botswana, Namibia, Zambia, and Zimbabwe. Another such treasure can be found in a section of Canada, with a section of land slipping into the United States. The Y2Y—Yellowstone to Yukon Conservation Initiative—traverses national and provincial parks and additional public lands to protect its wide-ranging species.

In 2016, Russia created the world's largest Arctic park outside of Greenland. Spanning the Franz Josef Archipelago with its 191 isles, and six times larger than the Serengeti, it's situated at the juncture of the Barents and Kara Seas. Polar bears, walrus, and vast sea bird colonies get a reprieve. On the same day in 2016, the United States similarly announced creation of a national marine protected area, this one larger than Texas. Its borders stretch from Hawaii to the Midway Islands.

Should people care about wildlife in vast ecosystems so distant that few will ever visit? Those who argue yes claim that ecosystems that offer clean air and healthy watersheds will ultimately benefit humans at local and global levels as well as wildlife. Others who favor these protected areas champion the potential economic gains when people do visit these areas. This economic argument might ring true in parts of Mongolia or Norway, but for places like Angola or northern Pakistan, where tourist numbers are understandably so limited, such logic carries little resonance.

Massive areas of Canada, and certainly Siberia and the Tibetan Plateau, will never be tourist meccas either, despite their unique and immensely biodiverse treasures. The plateau, in its own right, is not just a force in global weather and water levels but is also species rich, a richness matched by its high-elevation indigenous cultures. Though the Tibetan Plateau—the earth's only spot reflective of a what Beringia was like in the past—is dry, unrivaled, and with a bestiary like nowhere else, the logistics and money to get there, and access issues in general, are real considerations, raising the question of why anyone should support conservation in such insufficiently charted domains as these.

One reason is the scientific value of these areas, but that argument lacks luster locally. Whether such an area provides a needed standard on which to gauge ecological change or is merely the object of pure fascination, its value to science has little cachet with a Tibetan or Wyoming herder, an Arctic subsistence hunter, or an executive at Shell Petroleum. Local involvement is requisite in making decisions about the fate of undeveloped areas, but such tough choices also require engagement at the government level.

A more tempered view capitalizes on the value of such places simply existing: knowing that relatively unscathed kingdoms persist, even if not often visited, is considered important to many. People want to know elephants or orcas persist in situ, even if they, themselves, might only ever get to enjoy them in zoological parks.

An allied perspective comes through the lens of kindness, which is the recognition that we're all part of the same living system. In 2011, His Holiness the 17th Gyalwang Karmapa, Ogyen Trinley Dorje, wrote of Buddhism and conservation's inextricability: "No one thing exists by itself alone, or can survive alone. We are all part of one world ecology." And, in an interview four years later, he said: "Our share of this responsibility is to take what scientists teach us to heart, so we actually transform our way of life into one that is sustainable."

The faces of the world's *Nyimas* and *Tserings* come in many forms and cultures, as do belief systems and politics. Shedding species boundaries and distributing more generosity throughout the world could be a nice experiment.

The Dovrefjell Plateau in Norway's interior is high, cold, and windy. I'm hiking with Tord Breton, a government wildlife biologist. It's late summer, but snow has fallen. We cross permafrost and slide down chutes that have a glacier-like feel. The goal today is to find and necropsy a young muskoxen that has died.

Muskoxen struggle here. They did well initially, after their introduction into this deglaciated land where they had never trod before, but death is more common now, and an especially serious spate occurred in 2006. The dead had full bellies and lots of fat, so clearly they had not starved. Instead, their lungs showed signs of infection caused by bacteria consistent with pneumonia—something that had not previously affected the population from the time of their release here in the 1940s until the 2006 outbreak.

What had changed? Humidity on the Dovrefjell Plateau has increased 20 percent in the last few decades, as have temperatures. With such climate change comes the movement of pathogens into the area.

Farther north live the Sami, the herders of the Scandinavian Peninsula. Like other indigenous cultures, they capitalize on beasts of burden. Sami rely on reindeer, much like those in central Asia, say, have yaks or camels. Elsewhere, harvestable wild species such as salmon, gooseberries, or belugas still facilitate human survival. Such relationships are being expunged or inalienably altered as we confront climate modification and habitat loss, along with increasingly advanced modern amenities. For most of the world's species, our relationships with them are less direct, not utilitarian. Their survival, however, will also narrow in proportion to our self-indulgence.

In 1890, the year of the last Indian massacre in the United States at Wounded Knee in South Dakota, Crowfoot, a Lakota Sioux was asked, "What is life?"

"It is the flash of a firefly in the night. It is the breath of a buffalo in the wintertime. It is the little shadow which runs across the grass and loses itself in the sunset."

One hundred and twenty years later, I coax Freddy Goodhope into answering the same question.

"It is the earth, the ducks that fly over or the whales that swim by. It is the caribou that migrate."

He then asks me what life is. I suggest that it's the cycle of snow and the melting of ice, the deep tracks of a bear, a winter turning to spring, and the birth of baby muskoxen.

Postscript

A long time ago, northern realms slept in peace. Cold and ice blocked the peopling of a new world. With Earth's warming, adventuresome Asians crossed a land bridge to become the first Americans. With further melting, the bridge submerged.

Ice grasped Earth's high and cold extremes well before primates walked. Species reliant on the periglacial environment in the Andes, in high-elevation Asia, and at the world's northern edges are challenged. We colonize where ice sublimates. We always have. Our memories, like our lives, are short.

The two billion people living below Himalayan massifs are weather dependent. They rely on monsoonal moisture and glacial retention for their subsistence. Today's swift melt is not an asset, however, because it can result in flooding. Farther above, herders need healthy rangelands. So do and the yaks they herd as well as those in the wild, along with the other endemic species. Even in winter, female yaks are tied to snow because they need its moisture to produce milk for their still-nursing young. Glacial dissonance and the latest waves of humans anxious to colonize these rarified dominions occur at a pace that makes adjustments difficult.

The story differs little in the Far North, and only in kind. When the ice melts early, hunters cannot access rivers to find caribou or moose. Where ice recedes from the sea, Beringia's specialist animals confront more than warming: commercial vessels cross the top of the world—through Canada's infamous Northwest Passage and across Eurasia's lesser-known Northeast Passage, descending through the dangerous straits that once linked Asia and North America. While vast cargoes, petroleum products, and other goodies for humankind are transported through these shipping lanes—which also bring luxury cruises and tourists understandably keen to see the Arctic—we know little of their effect on the marine mammals that sustain the region's indigenous people, beyond that whales, seals, and walrus are sensitive to disturbance. We do know what a catastrophic oil spill will do.

While we also know that a postapocalyptic existence is possible for some animals on land and for some people, we don't know what that will be like for others.

The planet continues its transformation, one phase to another, and another.

Big remote places retain the fierceness of its wild animals and the thrill of unadulterated exploration, a quality that has vanished from much of the world. We alone have the capacity to retain a semblance of our biological heritage by keeping animals wild on the land. Is this necessarily a bad thing?

Individuals and entire societies covet mineral resources, including those from mountains and the steppes of central Asia and those found in the far north. Claims of sovereignty over mineral rights are reasonable, but not when avarice plays a role in the last of the wild melting away. Earth's warming causes geopolitical rancor to increase. In the interest of national security, Russia has expanded or entirely built eighteen military installations since 2014 in areas where the ice has been receding. Five are in Chukotka. China is armed with a new ice-breaking ship, though it has no polar claims.

Irrespective of nationalistic self-determinations, progress and optimism can be reconciled. Beyond Russia's newest national park, massive clean-up efforts continue on Wrangel. A new international Polar Code recognizes the value of the area and attempts to better protect the wildlife at the top of the world. Eight countries with Arctic holdings (Iceland is included here) have formal representation in discussion of these topics. In 2015, the U.S. Congress delegated an Arctic working group.

Beyond being gestures of good will, the gifting of animals by heads of state and the establishment of international peace parks can create opportunities to strengthen cooperation at scales that include joint management of species, scientific training, and detection of illicit trade. The public does take notice.

Still, the lesser known among the animal world await their turn. While kiangs and wild yaks are rebounding, on sanctified steppes and deep in the deserts, saigas, along with khulans and argalis, require a suppressing of the cashmere trade. Caribou simply need space, and muskoxen seek climate refugia.

I began this book with the serious question posed by Alexander Gruzdev. He asked whether geopolitics will affect the way we do science and the consequent message for conservation. The following is what I did not say, but it's what Sasha would understand.

The dictates of geography and history affect the realities of how science is done. Culture shapes which questions are asked, and how. These, in turn, form the silhouettes of what we'll achieve.

Science digs deeper. When there is no room in our hearts for gentleness, and when sympathy disappears from our vocabulary, so does conservation. It's then easy to imagine what path follows.

Acknowledgments

My folks, Alvin and Nathalie, told me to never stop dreaming. Parents across time have done similarly for their children.

During my formative postdoc years at Smithsonian's then Conservation and Research Centre, Chris Wemmer was an unwavering model. Because of Chris, along with John Seidensticker, the seeds of encountering far off lands and little known species took hold in me. George Schaller has always been an inspiration and admonished me not to let fieldwork escape my repertoire. He edited my first book and encouraged this one. Foremost though, it was Bill Weber and Amy Vedder who taught me that conservation without people is not conservation at all. Bill's favorite rebuke? "How is your work saving the world?" He was right to ask.

Passion at the Wildlife Conservation Society never relaxes. John Robinson has been an indomitable force for four decades. Ever since his days at the Smithsonian, he's been ardent in support of the boots that accompany science while internationalizing the conservation result. He's created the WCS that so many years ago I and others coveted to join. Jodi Hilty, the spirited leader of Yellowstone to Yukon Conservation Initiative is equally tireless. My graduate student nearly twenty years ago, and soon thereafter my boss at WCS, Jodi never said no—only what if and why not.

Former students and postdocs offer promise of successes across generations. John Goodrich introduced me to the Russian Far East and revealed why dedication to tigers makes a difference. Irina Goodrich used her native Russian interpretation skills to make collaborations on Wrangel Island a reality. Linda Kerley continues to champion Amur leopards. Jon Beckmann taught me about bears; we're now WCS colleagues. Janet Rachlow raised the science and conservation bar for Zimbabwe's white rhinos. Kevin White reaps insights about the migrations of Alaskan moose and mountain goats. Together, Kevin and I watched our first wild muskoxen near Greenland's ice cap. Julie Young added unflustered savvy to penetrate the dog-wildlife disharmony while helping with saigas. Matt Gompper and Jedidiah Brodie adopt bold insights on carnivores and creative ecology. Stefan Ekernas and Wesley Sarmento infiltrated the worlds of argalis and white goats, and Lishu Li that of ibex.

In Alaska, Janis Kozlowski and Katerina Wessels, of the National Park Service's Beringia's Heritage Program smoothed the way for my work in Beringia's wild east and west. Jim Lawler and Brad Schults helped with my first Arctic grant, taught me about muskoxen, and resolved logistical issues. Perry Barboza arranged photogrammetry visits to the Large Animal Research Station at the University of Alaska. Layne Adams brought the deep pockets of U.S. Geological Survey, an adroitness in handling muskoxen, and insights about Alaska's wildlife. Marci Johnson brought more than friendship—she solved serious aerial and ground logistics, shared data, and partook in talks in remote villages. Hillary Robison leads her own projects but manages to guide me.

In 1989, Terry Bowyer encouraged my first project in Alaska and each since. Gretchen Roffler, Wibke Peters, and Blake Lowrey served as my bio-techs in Alaska. Gretchen and Wibke now have their doctorates, and Blake is nearly there. Ross Schaefer Jr. and Lee Harris served as guides. Eric Sieh and Rick Swisher are the best of bush pilots anywhere.

Freddy Goodhope Jr. showed me reindeer and Arctic survival. He opened his home and his heart and proffered his unmatched knowledge of the Seward Peninsula. Also offering Alaskan village hospitality were Cliff and Eddie Weyhiouna and Leona Goodhope.

Linda Jeschke, Dan Stevenson, and the late Frank Hayes, on behalf of the National Park Service, smoothed my fieldwork beyond Kotzebue. Kimberlee Beckmann and Jim Dau helped collar six muskoxen. Scott Bergen assisted with spatial analyses. In Nome, Ken Adkisson is deep in his passion for subsistence harvest and muskoxen prehistory. Jeannette Pomraney facilitated public talks. Claudia Ihl always shared her insights from her long years of northern work.

Polar bear ecologists, especially Steve Amstrup, Karyn Rode, Andrew Derocher, Nikita Ovsyanikov, and Rocky Rockwell, helped make sense of the species they've studied for decades.

In Russia, Alexander Gruzdev remains a pillar for conservation. Not only does he handle massive logistical challenges but he has also been courageous in making Wrangel Island that international wonder that it is. Lizza Protas nimbly helped me try to understand Russians. Sergey Abarok was adept with translation skill, and together he and Lizza offered access to their splendid photos. Olga Starova helped with photogrammetry.

At Wrangel's clairvoyant weather station, Elena Victorovna, Kosper Alexandrovich, Sklrov (Igor) Georgievich, and Sergey Vladimirovich offered literal warmth and friendship. Field teams across two winter seasons who managed machines, observations, and miles of permafrost pleasure were Sanya Mikaevelavich, Igor Petrovich, Ilya Borisovich, Ivan Rostov, Sergei Stepanon, Gennnadiy Fedorov, and Gregoire (Grisha) Nikolaevich. Grisha and the two Igors and I all shared in a common language-less early morning calm. My Siberian colleagues assuredly have codified in their genomes the iced equivalent of cold-adaptiveness and alcohol tolerance that the DNA of Tibetans maintain for heightened hemoglobin.

In Mongolia, Amanda Fine facilitated the saiga project and workshops. Together with Peter Zahler, they make their conservation savvy count. Samantha Strindberg rendered the workshop that brought Mongolians to Montana successful. In artful attendance were Amgalan, Amgaa, Buuveibaatar, and Lhagvasuren. Buuvei has grasped the reins to understand Mongolia's ungulates. Fellow traveler Nara Urtnasan uses her language skills to make conservation tangible. Kirk Olson has devoted much of his life to conservation in Mongolia. Rich Reading has done similarly, and I have Rich to thank for my beginning in Asia. Ganchimeg Wingard gathered early economic data and later planted the first Mongolian flag in the Alaskan Arctic.

Among China's countless unsung heroes who made my fieldwork a reality were the untiring Aili Kang and Lishu Li, with help from Madhu Rao, Zhou Xin Lu, and Buqiong Buzhow. The forestry bureaus of Tibet and Qinghai and Kekexili National Nature Reserve were generous with permissions and other assistance. Drugay offered his magic and enthusiasm. Rich Harris and Andrew Smith were refreshingly honest, telling me when my ideas needed a better base in reality. Paul Buzzard kindly shared unpublished survey data.

Tshewang Wangchuk is among the world's most gifted conservationists and is unremitting in his commitment to Bhutan. Tempa Tshering and Tiger Sangay continue to ask what more can be done to enhance tiger and takin conservation. The Bhutan Foundation, Bhutan Council for Renewable Natural Resources, and Ugyen Wangchuk Institute and Center for the Environment, under the guidance of Nawang Norbu, sponsored my trips, as did Bruce Bunting. Kuenzang Phunso and Nancy Boedeker brought exciting knowledge and safety to the takin they so love.

Among colleagues and friends who have endured me with patience and shaped my career by their own science are Marc Bekoff,

Julie Betsch, Tord Breton, Ana Davidson, Erica Fleishman, Eric Dinerstein, Josh Donlan, Kathleen Galvin, Harry Greene, Jim Estes, Erin Johnson, Theresa Laverty, Gary Meffe, Clay Miller, Jed Murdoch, Mark and Delia Owens, Andrea Panagakis, Reed Noss, Michael Soule, Doug Smith, Peter Stacey, Ami Vitae, and Byron Weckworth. Within National Park Service and U.S. Fish and Wildlife Service circles is the ever gifted and untiring Elaine Leslie, who has shaped my career through migration cycles from local to global. Glen Plumb, Bert Frost, Gary Machlis, and Peter Dratch toil for America's wildlife. Charu Mishra wears less of a country hat and has helped understand the faces of both human and animal while imagining a better world.

At the University of Montana is Scott Mills, whose joint lab meetings enriched my years there, as did the amazing John Maron, Mark Hebblewhite, and Fred Allendorf. Mike Mitchell and Charlie Janson were my academic partners and advisers. Cindy Hartway used creativity to improve statistical insights and guide my projects. It was Dan Pletscher who never stopped trying to build the best program in wildlife biology anywhere.

Colorado State University has talented practitioners that blend the beauty of science with reality and challenge: George Wittemyer, Kevin Crooks, Chris Funk, John Hayes, Liba Pejchar, Sarah Reed, and never imitable Barry Noon. Ken Wilson has taken care of me for four years and is the quiet face behind Colorado State University's reigning success in conservation. Stewart Breck, who I once knew as a grad student twenty years earlier, invited me to share in his and Lauren's alpaca farm, while the U.S. Department of Agriculture's National Wildlife Research Center sponsored my Colorado a sabbatical.

The National Geographic Society has supported my work on and off since the 1980s. The Trust for Mutual Understanding has offered me help for excursions from Mongolia to the Asian half of Beringia. The National Park Service has done so across geographies from Washington, DC, and Glacier National Park to the Tetons and Alaska. Going further afield is the generosity of the Indianapolis Zoo, which under the guiding eyes of Mike Crowther and Karen Burns, recognizes world conservation leaders to better conserve wildlife across all continents and oceans. The Banovich Wildscapes Foundation has, likewise, gone beyond the mortal realm to recognize conservation leaders.

An organization is only as good as its people, and WCS continues with the best. Cristián Samper, as president and CEO, sets an impeccable standard. I especially appreciate that I was not let go after his talented partner, Adriana Casas, was seriously peppered with bear repellant under my tutelage. Those within WCS who inspire me and kept me afloat are Shannon Roberts, Kathryn Dunning, Keith Aune, Jon Beckmann, Liz Bennett, Jeff Burrell, Paul Calle, John Calvelli, Natalie Cash, Mary Dixon, Steve Fairchild, Josh Ginsburg, Julie Kunen, Heidi Kretser, Caleb McClellan, Sarah Reed, Steve Sautner, Scott Smith, Renee Seidler, Jon Slaght, Joe Walston, John Weaver, and Peter Zahler. Dale Miquelle and Justina Ray, outstanding colleagues for decades, abetted me during my little detention dance in Russia. Martin Robards has helped me across the years—from our first meetings in bush Arctic airports when I could not remember his name to now, when I can forget neither his name nor his efforts.

Paula White's generosity in time, planning, and advice is boundless. Her on-the-ground efforts from Africa to the Aleutians are a model for inspiration and why empiricism still matters.

Kent Redford and Steve Zack remain among my closest confidants, each who I knew before WCS, then as WCS colleagues, and now beyond. Together, we've laughed and cried, rafted Arctic rivers, explored other parts of the world, and written joint papers. Wherever they go, they instill in me and others a fascination for this planet's life.

At the University of Chicago Press, I'll never forgive Christie Henry, the senior editor of life sciences, who in 1991 gave me a volume on the Arctic. It began a love affair with Beringia that continues. Christie's perceptiveness and encouragement still guide me. Both Sheila Ann Dean and Yvonne Zipter offered thorough and critical edits of my initial manuscript. Each prodded in efficient ways to bring this book to fruition.

I owe special gratitude to Ellen Cheng, who suffuses compassionate conservation and matches her kindness with intellect. Steve Cain showed me that more can be done by silent valor. Susan Kutz offers passion for people and animals that showcases different pathways to success. Jon Swenson demonstrates a gentler approach to bringing data to the table. Joanna Lambert blends curiosity with intrepidness to understand animals. Sanjayan holds lessons for all. It's not that his international compass is excessively broad but his exemplariness for reaching the public where the science means less, but the conservation message more. David Quammen has done exactly this—but differently, only in the Quammen way—an incisive style and books that are the best that bio-journalism offers. Steve Zack and Bill Weber believed in me. Julie Kunen, now my exceptionally provident boss at WCS, Ana Davidson, now a CSU colleague, and Zack read earlier drafts of this manuscript. Katarzyna Nowak brought enthusiasm and wisdom during the home stretch. To all—I've been extraordinary fortunate to enjoy your company and learn from you as teachers, friends, and colleagues.

My small and intrepid family has always been there—Sonja, Taylor, Carol, Doris, Neal, and Bart.

Readings of Interest

Exploration

Andrews, R. C. 1932. *The New Conquest of Central Asia*. New York: American Museum of Natural History.

Baker, I. A. 2004. *The Heart of the World: A Journey to Tibet's Lost Paradise*. New York: Penguin Books.

Berger, J. 2008. *The Better to Eat You With: Fear in the Animal World*. Chicago: University of Chicago Press.

Burch, E. S., Jr. 2012. *Caribou Herds of Northwest Alaska, 1850–2000*. Fairbanks: University of Alaska Press.

Buzzard, P. J., et al. 2010. "A Globally Important Wild Yak *Bos mutus* Population in the Arjinshan Nature Reserve, Xinjiang, China." *Oryx* 44:577–80.

Crawford, R. M. 2013. *Tundra-Taiga Biology*. Oxford: Oxford University Press.

Cunningham, C., and J. Berger. 1997. *Horn of Darkness: Rhinos on the Edge*. New York: Oxford University Press.

Darwin, C. 1859. *On the Origin of Species*. New York: Collier Books.

Ehrlich, G. 2010. *In the Empire of Ice: Encounters in a Changing Landscape*. Washington, DC: National Geographic Books.

Estes, J. A. 2016. *Serendipity: An Ecologist's Quest to Understand Nature*. Berkeley: University of California Press.

Gray, D. 1987. *The Muskoxen of Polar Bear Pass.* Markham, Ontario: Fitzhenry & Whiteside.

Guthrie, R. D. 1990. *Frozen Fauna of the Mammoth Steppe: The Story of Blue Babe.* Chicago: University of Chicago Press.

Hansen, J., et al. 2009. "Black Soot and the Survival of Tibetan Glaciers." *Proceedings of the National Academy of Science* 106:22114–18.

Harris, R. B. 2007. *Wildlife Conservation in China: Preserving the Habitat of China's Wild West.* Armonk, NY: M. E. Sharpe.

Hedin, S. 1910. *Trans-Himalaya.* Vol. 1. New York: Macmillan.

———. 1922. *Southern Tibet.* Vol. 3. Stockholm: Lithograph Institute.

Hoffecker, J. F. 2005. *A Prehistory of the North: Human Settlement of the Higher Latitudes.* New Brunswick, NJ: Rutgers University Press.

Kingdon-Ward, F. 1913. *The Land of the Blue Poppy: Travels of a Naturalist in Eastern Tibet.* Cambridge: University Press.

Kinloch, A. 1892. *Large Game Shooting in Thibet, the Himalayas, Northern and Central India.* Calcutta: Thacker, Spink, and Company.

Klein, D. A. 1996. "Arctic Ungulates at the Northern Edge of Terrestrial Life." *Rangifer* 16:51–56.

Krupnik, I., and D. Jolly. 2002. *The Earth Is Faster Now.* Fairbanks: ARCUS.

Lent, P. C. 1999. *Muskoxen and Their Hunters.* Norman: University of Oklahoma Press.

Leslie, D. M., Jr., and G. B. Schaller. 2008. *Bos grunniens* and *Bos mutus* (Artiodactyla: Bovidae). *Mammalian Species* 836:1–17.

Marris, E., 2013. *Rambunctious Garden: Saving Nature in a Post-Wild World.* New York: Bloomsbury Publishing.

Matthiessen, P. 1978. *The Snow Leopard.* New York: Viking Press.

Mech, L. D., D. W. Smith, and D. R. MacNulty. 2015 *Wolves on the Hunt: The Behavior of Wolves Hunting Wild Prey.* Chicago: University of Chicago Press.

Miller, G. F. 1986. *Bering's Voyages: The Reports from Russia.* Fairbanks: University of Alaska Press.

Mingle, J. 2015. *Fire and Ice: Soot, Solidarity, and Survival on the Roof of the World.* New York: Macmillan.

Mithen, S. 2011. *After the Ice: A Global Human History, 20,000–5000 BC.* New York: Weidenfeld & Nicolson.

Niven, Jennifer. 2001. *The Ice Master: The Doomed 1913 Voyage of the Karluk.* New York: Hyperion.

O'Neill, D., 2009. *The Last Giant of Beringia: The Mystery of the Bering Land Bridge.* New York: Basic Books.

Phelps, E. 1900. "Yak Shooting in Tibet." *Journal of the Bombay Natural History Society* 13:134–43.

Pielou E. C. 1991. *After the Last Ice Age: The Return of Life to Glaciated North America.* Chicago: University of Chicago Press.

Post, E., et al. 2013. "Ecological Consequences of Sea-Ice Decline." *Science* 341: 519–24.

Quammen, D. 2004. *Monster of God: The Man-Eating Predator in the Jungles of History and the Mind*. New York: W. W. Norton & Co.

———. 2012. *Spillover: Animal Infections and the Next Human Pandemic*. New York: W. W. Norton & Co.

Prejevalsky, N. 1876. *Mongolia, the Tangut Country, and the Solitudes of Northern Tibet*. Vol. 1. London: Sampson, Low, Marston, Searle, and Rivington.

———. 1879. *From Kulja, across the Tian Shan to Lob-Nor*. Vol. 2. London: Sampson, Low, Marston, Searle, and Rivington.

Rasmussen, K. 1927. *Across Arctic America*. New York: G. P. Putnam & Sons.

Rawling, C. G. 1905. *The Great Plateau: Being an Account of Exploration in Central Tibet, 1903, and of the Gartok Expedition, 1904–1905*. London: Edward Arnold.

Roberts, D. 2005. *Four against the Arctic: Shipwrecked for Six Years at the Top of the World*. New York: Simon and Schuster.

Rockhill, W. W. 1895. "Big Game of Mongolia and Tibet." In *Hunting in Many Lands*, ed. T. Roosevelt and G. G. Bird, 255–77. New York: Forest and Stream.

Schaller, G. B. 1978. *Mountain Monarchs: Wild Sheep and Goats of the Himalaya*. Chicago: University of Chicago Press.

———. 1982. *Stones of Silence: Journeys in the Himalaya*. New York: Bantam Books.

———. 1998. *Wildlife of the Tibetan Steppe*. Chicago: University of Chicago Press.

———. 2012. *Wild Tibet: A Naturalist's Journeys on the Roof of the World*. Washington, DC: Island Press.

Schaller, G. B., A. Kang, T. D. Hashi, and P. Cai. 2007. "A Winter Wildlife Survey in the Northern Qiangtang of Tibet Autonomous Region and Qinghai Province, China." *Acta Theriologica Sinica* 27:309–16.

Vibe, C. 1967. *Arctic Animals in Relation to Climatic Fluctuations*. Meddelelser om Groenland, vol. 170. Copenhagen: C. A. Reitzel.

Wallace, A. R. 1878. *Tropical Nature, and Other Essays*. London: Macmillan Co.

Wellby, M. S. 1898. *Through Unknown Tibet*. London: Fisher, Unwin.

Wilson, M. C., and A. T. Smith. 2015. "The Pika and the Watershed: The Impact of Small Mammal Poisoning on the Ecohydrology of the Qinghai-Tibetan Plateau." *Ambio* 44:16–22.

Wit, P., and I. Bouman. 2012. *The Tale of the Przewalski's Horse: Coming Home to Mongolia*. Amsterdam: KNNV Publishing.

Xu, J., et al. 2009. "The Melting Himalayas: Cascading Effects of Climate Change on Water, Biodiversity, and Livelihoods." *Conservation Biology* 23:520–30.

This Project

Batsaikhan, N., et al. 2014. "Conserving the World's Finest Grassland amidst Ambitious National Development." *Conservation Biology* 28:1736–73. DOI: 10.1111/cobi.12297.

Berger, J. 1997. "Population Constraints Associated with Black Rhinos as an Umbrella Species for Desert Herbivores." *Conservation Biology* 11:69–78.

———. 2003. "Is It Acceptable to Let a Species Go Extinct in a National Park?" *Conservation Biology* 17:1451–54.

———. 2004. "The Longest Mile: How to Sustain Long Distance Migration in Mammals." *Conservation Biology* 18:320–32.

———. 2007a. "Carnivore Repatriation and Holarctic Prey: Narrowing the Deficit in Ecological Effectiveness." *Conservation Biology* 21:1105–16.

———. 2007b. "Fear, Human Shields, and the Re-Distribution of Prey and Predators in Protected Areas." *Biology Letters* 3:620–23.

———. 2008a. *The Better to Eat You With: Fear in the Animal World*. Chicago: University of Chicago Press.

———. 2008b. "Undetected Species Losses, Food Webs, and Ecological Baselines: A Cautionary Tale from Yellowstone." *Oryx* 42:139–43 and 42:176.

———. 2012. "Estimation of Body-Size Traits by Photogrammetry in Large Mammals to Inform Conservation." *Conservation Biology* 26:769–77.

———. 2013. "Ignoring Nature: Failing to Learn or Learning to Fail?" In *Ignoring Nature*, ed. M. Bekoff, 113–17. Chicago: University of Chicago Press.

———. 2017. "The Science and Challenges of Conserving Large Wild Mammals in 21st-Century American Protected Areas." In *Science, Conservation, and National Parks*, ed. S. Beissinger et al., 188–211. Chicago: University of Chicago Press.

Berger, J., B. Buuveibaatar, and C. Mishra. 2013. "Globalization of the Cashmere Market and the Decline of Large Mammals in Central Asia." *Conservation Biology* 27:679–89.

Berger, J., and S. L. Cain. 2014. "Moving beyond Science to Protect a Mammalian Migration Corridor." *Conservation Biology* 28, 1142–50.

Berger, J., E. Cheng, M. Krebs, L. Li, A. Kang, G. B. Schaller, and M. Hebblewhite. 2015. "Legacies of Past Exploitation and Climate Affect Mammalian Sexes Differently on the Roof of the World—the Case of Wild Yaks." *Scientific Reports—Nature*. 5:8676. DOI: 10.1038/srep08676.

Berger, J., and C. Cunningham. 1994. "Active Intervention and Conservation: Africa's Pachyderm Problem." *Science* 263:1241–42.

Berger, J., C. Hartway, A. Gruzdev, and M. Johnson. 2018. "Climate Degradation and Extreme Icing Events Constrain Life in Cold-Adapted Mammals." *Scientific Reports* 8:1156. DOI: 10.1038/s41598-018-19416-9.

Berger, J., J. E. Swenson, and I.-L. Persson. 2001. "Recolonizing Carnivores and Naive Prey: Conservation Lessons from Pleistocene Extinctions." *Science* 291:1036–39.

Berger, J., et al. 2001. "Back-Casting Sociality in Extinct Species: New Perspectives Using Mass Death Assemblages and Sex Ratios." *Proceedings of the Royal Society* 268:131–39.

Berger, J., et al. 2006. "Connecting the Dots: An Invariant Migration Corridor Links the Holocene to the Present." *Biology Letters* 2:528–31.

Berger, J., et al. 2008. "Protecting Migration Corridors: Challenges and Optimism for Mongolian Saiga." *PLoS Biology* 6:166–68.

Berger, J. et al., 2010. "Captures of Ungulates in Central Asia Using Drive Nets: Advantages and Pitfalls as Illustrated by Endangered Mongolian Saiga." *Oryx* 44:512–15.

Berger, J., et al. 2014. "Sex Differences in Ecology of Wild Yaks at High Elevation in the Kekexili Reserve, Tibetan Qinghai Plateau, China." *Journal of Mammalogy* 95: 638–45.

Buuveibaatar, B., J. K. Young, J. Berger, et al. 2013. "Factors Affecting Survival and Cause-Specific Mortality of Saiga Calves in Mongolia." *Journal of Mammalogy* 94: 127–36.

Ekernas, S., W. M. Sarmento, H. S. Davie, R. P. Reading, J. Murdoch, G. J. Wingard, S. Amgalanbaatar, and J. Berger. 2017. "Desert Pastoralists' Negative and Positive Effects on Rare Wildlife in the Gobi." Conservation Biology 31:269-77. DOI: 10.1111/cobi.12881.

Estes, J. A, et al. 2011. "Trophic Downgrading of Planet Earth." *Science* 333:301–6.

Sarmento, W., and J. Berger. 2017. "Human Visitation Limits the Utility of Protected Areas as Ecological Baselines." *Biological Conservation* 212:326–36. http://dx.doi.org/10.1016/j.biocon.2017.06.032.

Young, J., et al. 2010. "Population Estimates of Mongolian Saiga: Implications for Effective Monitoring and Population Recovery." *Oryx* 44:285–92.

Young, J., et al. 2011. "Is Wildlife Going to the Dogs? A Review of the Impacts of Feral and Free-Roaming Dogs on Wildlife Populations." *Bioscience* 61:125–32.

Op-Eds and Blogs

Angier, N. 2010. "Musk Oxen Live to Tell a Survivor's Tale." *New York Times*, December 14, http://www.nytimes.com/2010/12/14/science/14angier.html?_r=0/.

Berger, J. 2012. "Arctic Alaska's Conservation Conundrum." *Voices* (blog), *National Geographic*, February 2. http://voices.nationalgeographic.com/2012/02/02/arctic-alaskas-conservation-conundrum/.

Berger, J. 2013. "The Damage Cashmere Does." *New York Daily News*, September 9 http://www.nydailynews.com/opinion/damage-cashmere-article-1.1450204.

Berger, J. 2014a. "On Far-Flung Wrangel Island, a Scientist Sizes up Muskoxen." *Yale-Environment360*, April 2. https://e360.yale.edu/digest/on_far-flung_wrangel_island_a_scientist_sizes_up_muskoxen/4121/.

Berger, J. 2014b. "Studying a Polar Menagerie on an Island in Arctic Russia." *Yale-Environment360*, April 16. http://e360.yale.edu/digest/studying_a_polar_menagerie_on_an_island_in_arctic_russia/4126/.

Berger, J. 2014c. "Russian-American Collaboration Carries on in Key Arctic Ecosystem." *YaleEnvironment360*, April 30. http://e360.yale.edu/digest/russian-american_collaboration_carries_on_in_key_arctic_ecosystem.

Berger, J. 2014d. "Polar Bear Diplomacy: Where the US and Russia Can Agree." *Christian Science Monitor Weekly*, May 21, http://www.csmonitor.com/Commentary/Opinion/2014/0521/Polar-bear-diplomacy-Where-the-US-and-Russia-can-agree.

Berger, J. 2016a. "The Unlikely Diplomats: The Return of Muskoxen to Alaska Marks at Least One Success for U.S.–Russia Relations." *US News & World Report*, January 8. http://www.usnews.com/opinion/blogs/world-report/articles/2016-01-08/the-hairy-arctic-mammal-that-thawed-us-russia-relations.

Berger, J. 2016b. "Silent Soldiers of the Extreme, or Why I'm Glad I'm Not a Wild Yak." Mongabay, November 22. http://www.news.mongabay.com/2016/11/silent-soldiers-of-the-extreme-or-why-im-glad-im-not-a-wild-yak/.

Berger, J. 2017. "Scientist at Work: Tracking Muskoxen in a Warming Arctic." *The Conversation*, February 12. https://theconversation.com/scientist-at-work-tracking-muskoxen-in-a-warming-arctic-70378.

Gaworecki, M. 2017. "Newcast#9: Joel Berger on Overlooked 'Edge Species' That Deserve Conservation." January 10. In *Almost Famous Animals*, podcast. Mongabay. https://news.mongabay.com/2017/01/newscast-9-joel-berger-on-overlooked-edge-species-that-deserve-conservation/.

Groskin, L. 2017. "Bear in Mind the Muskoxen." May 17, in *Science Friday*, produced by Luke Groskin. Public Radio International, interview and video, http://sciencefriday.com/videos/bear-in-mind-the-muskox/.

Lee, M. 2013. "A Life in the Wild—Biologist Joel Berger." [In Chinese.] *Outside*, China edition, January 24, 124–27.

Ng, D., and J. Berger. 2017. "The Hidden Cost of Cashmere." *Inside Asia* (blog), *Forbes*, February 16. http://www.forbes.com/sites/insideasia/2017/02/16/cashmere-cost-environment/#21458cfa4378.

Ng, D., and J. Berger. 2017. "Dogs Are Man's Best Friend—But One of Wildlife's Worst." *Washington Post*, March 24. https://www.washingtonpost.com/opinions/dogs-are-mans-best-friend--but-one-of-wildlifes-worst-foes/2017/03/24/2d2c9232-ed7f-11e6-9973-c5efb7ccfb0d_story.html?utm_term=.2f5e9097ea4f.

Wei-Haas, M. 2017. "Special Report: Future of Conservation: To Understand the Elusive Musk Ox, Researchers Must Become Its Worst Fear." *Smithsonian*, August 2. http://www.smithsonianmag.com/science-nature/understand-elusive-muskox-researchers-must-become-creatures-worst-fear-180964287/.

Zimmer, C. 2018. "In the Arctic, More Rain May Mean Fewer Musk Oxen." *New York Times*, January 18. https://www.nytimes.com/2018/01/18/science/musk-oxen-climate-change.html?_r=0.

Index